Trace fossils from the Lower Cambrian Mickwitzia s south-central Sweden

SÖREN JENSEN

Jensen, S. 1997 03 26: Trace fossils from the Lower Cambrian Mickwitzia sandstone, south-central Sweden. *Fossils and Strata*, Nr. 42, pp. 1–110. Oslo. ISSN 0300-9491. ISBN 82-00-37665-6

The Mickwitzia sandstone, south-central Sweden, consists of about 10 m of Lower Cambrian clastic sediments deposited in an epicontinental setting. An informal, lithologically based subdivision, A–E, is introduced. A thin basal conglomerate (interval A) is followed by thin-bedded sand and siltstone with clayey partings (interval B and D) and medium-grained sandstone (interval C), largely representing subtidal storm deposits. Interval E consists of thick-bedded shoreface deposits. Heterolithic intervals have well-preserved trace fossils, including *Cruziana*, *Rusophycus*, *Gyrolithes*, *Treptichnus* and *Teichichnus*. Beds with impure, often weakly cemented sandstone (interval C) have *Rhizocorallium*, *Monocraterion* and *Skolithos*. Trace fossils are dominated by infaunal feeding and feeding?–dwelling burrows; 40 ichnotaxa are recognized, representing the activity of but a few types of animals. The type material of *Monocraterion tentaculatum* Torell, 1870, is illustrated for the first time, and the relationship of *Monocraterion* to *Skolithos* and *Rosselia* is discussed. Previously poorly known taxa are described. *Scotolithus mirabilis* Linnarsson, 1871, consists of a vertical shaft which in its lower part diverges into a wide broom-shaped arrangement. *Spiroscolex spiralis* (Torell, 1870) is a little-used name for burrows identical to *Gyrolithes polonicus*. *Halopoa imbricata* Torell, 1870, is a burrow related to *Palaeophycus sulcatus*, with a morphology dependant on sediment consistency: it is here assigned to *Palaeophycus imbricatus*. *Fraena tenella* Linnarsson, 1871, is assigned to *Cruziana* and considered a subjective senior synonym of *Cruziana problematica*. *Phycodes pedum* Seilacher, 1955, should be assigned to *Treptichnus*. □*Trace fossils, Lower Cambrian, Mickwitzia sandstone, Sweden.*

Sören Jensen [sj10019@rock.esc.cam.ac.uk], Department of Geosciences, Historical Geology and Palaeontology, Norbyvägen 22, S-752 36 Uppsala, Sweden; current address: Department of Earth Sciences, University of Cambridge, Downing Street, Cambridge CB2 3EQ, UK; 8th October, 1993; revised 1st November, 1995.

Contents

Introduction

The Mickwitzia sandstone, in older literature called the Eophyton sandstone, forms the basal part of the Cambrian sedimentary succession in Västergötland, south-central Sweden. The Mickwitzia sandstone is a mostly thin-bedded siltstone and sandstone with clayey interbeds rarely exceeding a few centimetres in thickness. The Mickwitzia and overlying Lingulid sandstones constitute western members of the File Haidar Formation, a dominantly siliclastic succession named after its type area on Gotland, eastern Sweden (Thorslund & Westergård 1938; Bergström & Gee 1985; Hagenfeldt 1994) (Fig. 1). In Västergötland the File Haidar Formation reaches a thickness of about 35 m, of which the Mickwitzia sandstone member makes up about 10 m. Impoverished in body fossils, the Mickwitzia sandstone is best known for abundant well-preserved trace fossils, sedimentary structures, and

problematic structures (e.g., Torell 1868, 1870; Linnarsson 1869a, 1871; Nathorst 1874, 1886a; Westergård 1931a; Bergström 1968, 1973a; Jensen 1990). Thus, *Spatangopsis* and *Protolyellia* gained fame in the late 19th century by being interpreted as remnants of medusae (Nathorst 1881a, 1910) and, more recently through Seilacher's (e.g., 1983a, 1984, 1992a, 1994a) interpretation of the structures as sand-skeletons of cnidarians. In the debate over whether 'fucoids' were the remains of plants or structures caused by animals and purely physical processes, Nathorst (1874, 1881b, 1886a), a strong proponent of the latter standpoint, made frequent reference to trace fossils from the Mickwitzia sandstone.

There is a considerable body of work dealing with different aspects of the Mickwitzia sandstone (e.g., Wallin 1868; Torell 1868, 1870; Linnarsson 1869a, b, 1871; Nath-

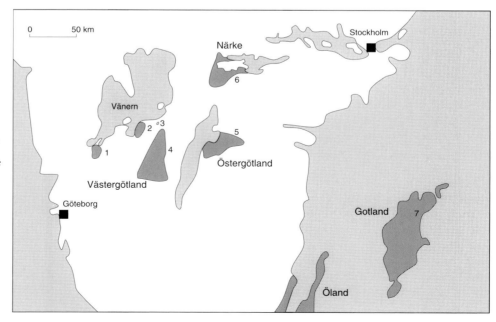

Fig. 1. Map of central Sweden showing principal areas of sedimentary cover (darker grey) where the File Haidar Formation occurs. 1, Halleberg and Hunneberg; 2, Kinnekulle; 3, Lugnås; 4, the Falbygden area, including Billingen; 5, Östergötland, with sedimentary inlier close to lake Vättern, including Bårstad 2 core; 6, Närke, with sediments preserved within a faulted area on western half of Lake Hjälmaren; 7, Gotland (subsurface), with number positioned at location of the File Haidar core.

orst 1874, 1881a, b, 1885, 1886a, b, 1910; Holm 1901; Moberg 1911; Hadding 1927, 1929; Westergård 1931a, b, 1943; Størmer 1956; Bergström 1968, 1973a; Martinsson 1974; Lindström 1977, 1984; Lindström & Vortisch 1978; Scholz 1977; Möller 1987; Jensen 1990). The purpose of this contribution is to present the first detailed examination of the trace fossils in the Mickwitzia sandstone and to give an interpretation of the environment in which they formed.

Material

Field work for this study was carried out at Lugnås, Kinnekulle and Billingen. A core from Kinnekulle (Mossen-1) was studied at the Geological Institute, Stockholm University. Most of the material studied came from collections of the Swedish Museum of Natural History, Stockholm (RM), the Swedish Geological Survey, Uppsala (SGU), the Palaeontological Museum, Uppsala (PMU), and the Department of Historical Geology, University of Lund (LO). This material consists largely of collections made during the later part of the 19th century by J.G.O. Linnarsson, G. Holm and, especially, the professional collector G. von Schmalensee. As a rule, this material lacks detailed stratigraphic and geographic information. An important collection of fossils from the basal parts of the Mickwitzia sandstone has been assembled from Hjälmsäter during several decades by Mr. Allan Karlsson, Hjälmsäter, the major part of which is now deposited at RM. A significant private collection has also been made by Mr. Jan Johansson, Sköllersta, Närke; material was also examined in the collection of Mr. Holger Buentke, Lugnås.

Previous work

Early students of the Lower Cambrian sandstone in Västergötland (e.g., Kalm 1746; Linnaeus 1747) noticed its position at the base of the sedimentary sequence, but the earliest stratigraphic subdivision of sedimentary rocks in Sweden was made by Angelin (1854), who, using Kinnekulle as a key region, defined five divisions, based mainly on the trilobites. The basal sandstone division was characterized by the lack of trilobites and the occurrence of supposed plant petrificates, and hence was named Regio Fucoidarum. The first to report fossils from the sandstone was Brongniart (1828), who identified and illustrated supposed plant remains as *Fucoides circinatus* Brongniart, 1828. Linnarsson (1866) reported impressions of algae and cylindrical tubes belonging to *Scolithus* (= *Skolithos*), and he later reported a brachiopod from Billingen, identified as *Lingula* (Linnarsson 1868). Up to this point the

sandstone had been treated as a single unit. Though the Mickwitzia sandstone was accessible, descriptions of rock character and fossils are based upon finds in the Lingulid sandstone.

The first major contribution to the knowledge of the Mickwitzia sandstone was made by Wallin (1868). He clarified the nature of the boundary with the underlying basement and introduced a twofold division of the sandstone. In particular, he studied the contact between the Precambrian gneiss and the sandstone at Lugnås, where it was readily accessible in several millstone quarries and mines; the rock used for millstone fabrication had variously been considered to be an arkose or to be part of the basement rock. Wallin established that the layer used for millstone fabrication was weathered gneiss. Separating the weathered gneiss from the main sandstone is a conglomeratic layer which Wallin treated as a separate unit, not included in the sandstone. Based upon general petrographic character, Wallin divided the sandstone into two units: the lower, Eophyton sandstone, was characterized as hard and with shaly intercalations; the upper, Fucoid sandstone, was characterized as less well-lithified. This division is still in use, although, following Holm (1901), the names have been changed to the Mickwitzia and Lingulid sandstone based on their contained brachiopods. The Mickwitzia sandstone as defined by Holm (1901) includes also the basal conglomerate, as recommended by Linnarsson (1869a).

Differing lithological characters and bedding thicknesses have continued to be the main distinguishing features of the two sandstones. The Mickwitzia sandstone is mostly thin-bedded, consisting of silt- and sandstones alternating with clayey layers, while the Lingulid sandstone is a more homogenous and fine-grained sandstone (e.g., Hadding 1929; Westergård 1931b). The former is hard, rich in mica-flakes, and often has haematitic staining, while the latter is cleaner and poorly cemented, with a higher porosity. However, no consistent picture has emerged on the transition between the Mickwitzia and Lingulid sandstones. Linnarsson (1871, p. 4) reported a friable conglomerate separating the Mickwitzia and Lingulid sandstone at Lugnås and Billingen. Westergård (1931b) did not observe a conglomerate but noticed a clear boundary where thin beds give way to thicker beds, marked by coarse-grained sediment and weathered feldspar not found in the immediately over- or underlying rocks. Holst (1893, p. 16) reported a varying thickness of the Mickwitzia sandstone at Lugnås, adding that lenses of Mickwitzia sandstone character occur about two metres above the base of the Lingulid sandstone. At Kinnekulle the transition between the two units has been described as gradational (Holm 1901; Westergård 1943). The total thickness of sandstone is given as 34 m, of which the Mickwitzia sandstone compromises the basal 8–10 m (e.g., Hadding 1929).

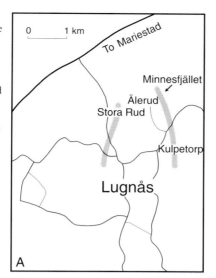

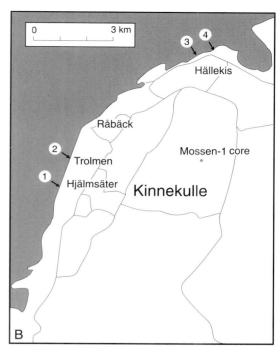

Fig. 2. Principal localities of the Mickwitzia sandstone at Kinnekulle and Lugnås. □A. Lugnås; Shaded pattern shows areas with millstone quarries and mines, most of which are now collapsed or overgrown. Arrow indicates location of restored mine at Älerud, now a cultural memorial, Minnesfjället. □B. Kinnekulle. 1, Shore at Hjälmsäter; 2, shore at Trolmen harbour; 3, shore at Hällekis; 4, small outcrop close to shooting range at Hällekis. Also marked is location for the Mossen core.

The first palaeontological studies on the Mickwitzia sandstone (Linnarsson 1869a, 1871; Torell 1868, 1870) identified well over twenty species as remnants of plants, worms and echinoderms. Nathorst's (1881a, 1910) work is pioneering for the understanding of trace fossils, and his monograph, 'Om spår af några evertebrerade djur m.m och deras paleontologiska betydelse' (Nathorst 1881b), is still worthwhile reading. Nathorst performed the precursors of neoichnological experiments by recording traces made by various organisms placed in plaster-of-paris and mud, comparing them with trace fossils, including several of those in the Mickwitzia sandstone.

Subsequent work on the fossils in the Mickwitzia sandstone has been scattered. Molds of the arthropod *Paleomerus hamiltoni* Størmer, 1956, are known from erratic boulders and generally thought to derive from the Mickwitzia sandstone (Størmer 1956; Bergström 1968, 1971), although this has recently been questioned (Möller 1987). Westergård's (1931a) study on *Diplocraterion, Monocraterion* and *Skolithos* includes several specimens from the Mickwitzia sandstone. Arthropod burrows from the Mickwitzia sandstone have been discussed by Bergström (1973a) and Jensen (1990). *Spatangopsis* and *Protolyellia*, previously interpreted as medusae (Nathorst 1881a, 1910), have recently been reinterpreted as sand skeletons (Seilacher, e.g., 1992a, 1994a) or holdfasts (Bergström 1989, 1991).

The Mickwitzia sandstone has long been interpreted as a shallow-water deposit formed as the sea transgressed a peneplained sub-Cambrian basement (Holm 1901; Hadding 1927, 1929). Ripple marks, alternating sand/clay layering, as well as purported mud cracks and rain-drop imprints were considered indicators of shallow water. Eklund (1961) considered the Mickwitzia sandstone in Närke to be a tidal deposit, while Martinsson (1974) invoked storm deposition.

An early Cambrian age for the Mickwitzia sandstone was suggested by Linnarsson (1869a, b, 1871) and Torell (1868, 1870). In the Baltic area the Mickwitzia sandstone was correlated with the 'Eophyton sandstone' in Estonia (Linnarsson 1874; Schmidt 1888; Öpik 1925), the 'zone with *Schmidtiellus Torelli*' in Scania (Moberg 1911) and the 'Zone with *Volborthella tenuis* or *Platysolenites antiquissimus*' in the Mjøsa area, Norway (Vogt 1924). Recently, more detailed biostratigraphic correlation of the Lower Cambrian of the Baltoscandian area has been achieved using body fossils (Bergström 1981; Bergström & Gee 1985; Ahlberg 1984) and acritarchs (Vidal 1981a, b; Moczydłowska & Vidal 1986; Moczydłowska 1991; Eklund 1990; Hagenfeldt 1989, 1994). This has placed the Mickwitzia sandstone in the upper part of the Lower Cambrian, within the Holmia kjerulfi Assemblage Zone.

Localities

The positions of localities are given according to the UTM grid of the topographical map of Sweden (*Topografisk karta över Sverige*).

Outcrops of the Mickwitzia sandstone in Västergötland are today restricted to the small hill Lugnås (Fig. 2A) and the western side of Kinnekulle (Fig. 2B). On the northern-most part of Billingen, (Fig. 1) quarries through parts of the Mickwitzia sandstone have been accessible at Prästorp and Stolan (Wallin 1868; Linnarsson 1869a, b) (UTM VE 30405857845), though now only loose material can be obtained. On Kinnekulle, patchy exposures are accessible on the shore of lake Vänern from Hällekis to Hjälmsäter (Fig. 2B). The basal part, showing contact of weathered gneiss to conglomerate and a few basal sandstone beds, is accessible at Trolmen harbour (UTM VE 0345096395) and just north of Råbäck harbour. Along the shore from Trolmen harbour south to Hjälmsäter, the strata form a gentle trough with beds of up to approximately the 1.4 m level exposed at the water level. About 700 m south of Trolmen harbour is a narrow man-made trench exposing all but the lowest 1.4 m (UTM VE 0319595495). Thick beds at the top of the Mickwitzia sandstone (often considered to be at the base of the Lingulid sandstone; e.g., Holm 1901; Westergård 1931a, b) form escarpments along much of the shore from Hjälmsäter to Hällekis. Only rare patches of lower levels are exposed in the scree slopes beneath these beds. At Hällekis, on the northern part of Kinnekulle, the highest few metres of the Mickwitzia sandstone and the gradational contact with the Lingulid sandstone are exposed near a shooting-range (UTM VF 0851500645), and 600 m south of this are exposures of beds about 6 m above the basal conglomerate (UTM VF 0814500580).

At Lugnås, blocks of Mickwitzia sandstone are found near the numerous abandoned millstone quarries, although outcrops are now rare. The major exception is a restored millstone mine, Minnesfjället, which is protected as a cultural memorial (UTM VE 2667599590). In this and the surrounding area the lower 7 m of the Mickwitzia sandstone are visible.

Profile description and interpretation

The following general description of the lithology and the physical and biogenic sedimentary structures of the Mick-witzia sandstone in Västergötland is based largely on observations from three sites:

1 *The Mossen-1 core.* – Core drilled in 1984 by SKB (Svensk kärnbränslehantering), housed at the Geological Institute in Stockholm has a complete or nearly complete sequence of the Mickwitzia sandstone (Figs. 2B, 3)
2 *Hjälmsäter–Trolmen.* – A profile exposing all but the lowermost metre of the Mickwitzia sandstone is accessible at Hjälmsäter, immediately adjacent to the shore of lake Vänern (Fig. 2B). The base of the profile consists of two thick beds of sandstone that can be followed northwards along the shore to Trolmen harbour

about 700 m north of the Hjälmsäter profile, where the lowermost metre is also accessible. At Hjälmsäter the horizontal extent of exposure is very limited, generally less than 2 m. This profile has been figured and briefly discussed by Scholz (1977) and in more detail by Möller (1987).
3 *Lugnås.* – At Lugnås, the lower 6–7 m were studied at and near a restored millstone mine at Älerud (Fig. 2A)

Examination of these three profiles revealed a similar lithological succession. Based on the Mossen-1 core, a lithological division A–E is recognized for convenience of reference (Fig. 3). The description combines information from the three sections, pointing out cases of known major differences. It also includes information from the literature. Measurements are taken from the core at Mossen. The vertical extent of the intervals differs slightly between the different sections (Figs. 3, 4). Only a number of petrographic analyses have been made for this study. Dominant grain size in the drill core was obtained by observation with a hand lens.

Interval A, 0–0.3 m

Description. – Basal conglomerate, polymict, matrix-supported. It rests on a denuded Precambrian gneiss basement of low relief. The development of the conglomerate varies; it may rest directly on the basement, be separated from it by sandstone, or be absent (Hadding 1927). At Trolmen harbour there is a basal bed about 5–15 cm thick with a rippled sandy upper part. The ripples have a wavelength of about 10 cm with a ripple index of 1, and strike from NW–SE. This bed is locally divided into two units. A clayey layer with a thickness of a few centimetres separates the lowermost part from an upper part, about 14 cm thick, which in places is developed as a pure sandstone. The latter bed has sinuous ripple crests, in places developed into an interference pattern, striking NE–SW, and cross-bedding that suggests currents from the southeast.

The pebbles in the conglomerate are composed of quartz and to a lesser extent feldspar, with a matrix of mostly medium- to coarse-grained sand and with a cement of quartz or calcite (Hadding 1927). Other components include fragments of sedimentary rocks (Hadding 1927). Especially at Lugnås, the conglomerate contains large pebbles of quartz, including some that have been compared to wind-faceted stones (Nathorst 1885, 1886b; Hadding 1927). At Lugnås, Westergård (1931b) reported that the conglomerate contained pebbles rich in siderite and mica. This type of rock occurs also in the conglomerate just south of Hjälmsäter but was not found at Trolmen. Examination of such pebbles from Lugnås shows them to be elongated, rounded clasts of fine-grained material, clay and silt, with occasional larger

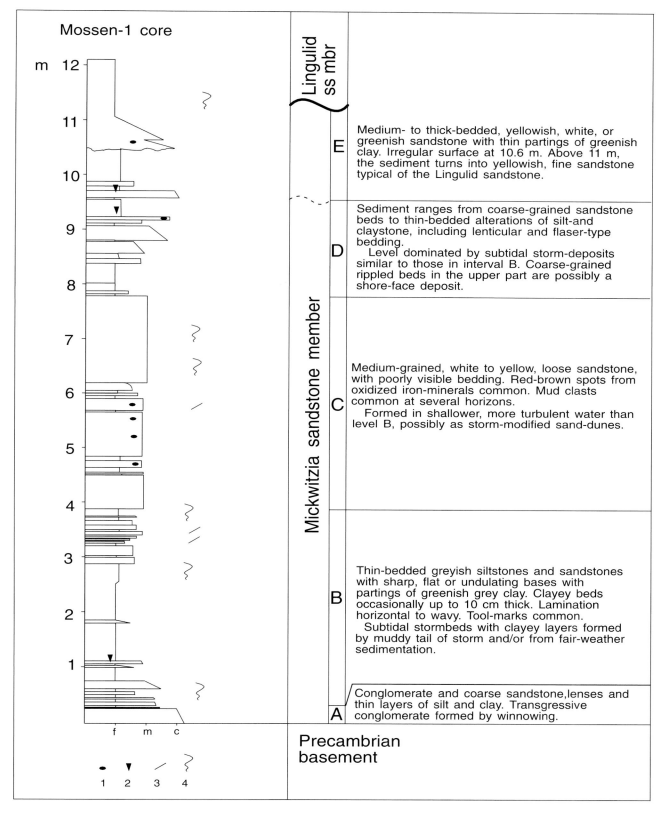

Fig. 3. Graphic log through the Mickwitzia sandstone, based on the Mossen core, Kinnekulle. For descriptive purposes, an informal division into intervals A–E has been used. Broken horizontal wavy line shows traditional boundary between the Mickwitzia and Lingulid sandstone. 1, Mud clasts; 2, sandstone-filled shrinkage cracks; 3, strongly uneven bedding contacts; 4, intensive bioturbation.

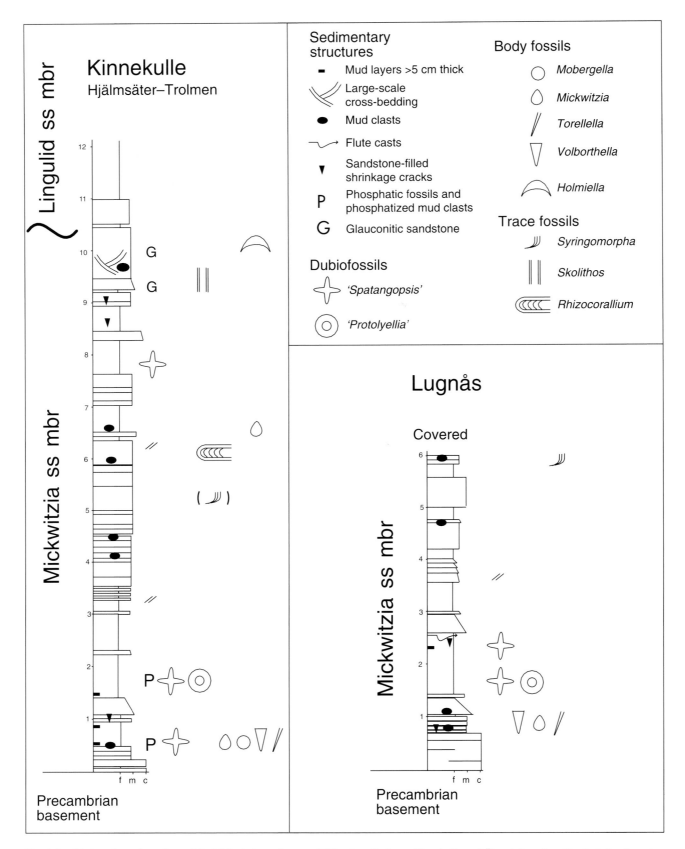

Fig. 4. Graphic logs through sections of the Mickwitzia sandstone at Hjälmsäter–Trolmen, Kinnekulle, and Älerud, Lugnås, with selected sedimentary structures, dubiofossils, trace fossil and body fossil.

Fig. 5. Sedimentary structures from the Mickwitzia sandstone. Scale bars 10 mm. □A. Filled shrinkage cracks occurring together with filled *Skolithos* tubes. Råbäck. Level E. RM X3289. □ B, C. Conglomeratic bed seen in side (B) and basal (C) view. Oval black structures are mud-clasts with siderite cement. Note incomplete shrinkage cracks at base of bed. Älerud, Lugnås. Identical blocks observed in upper part of level A at Älerud, Lugnås. RM X3290. □D. Filled irregular cracks at base of bed. Lugnås, SGU 8618. □E. Ridge pattern on lower surface of sandstone bed, possibly identical to 'Archae-orhizza' of Torell (1870). Lugnås. SGU 8577.

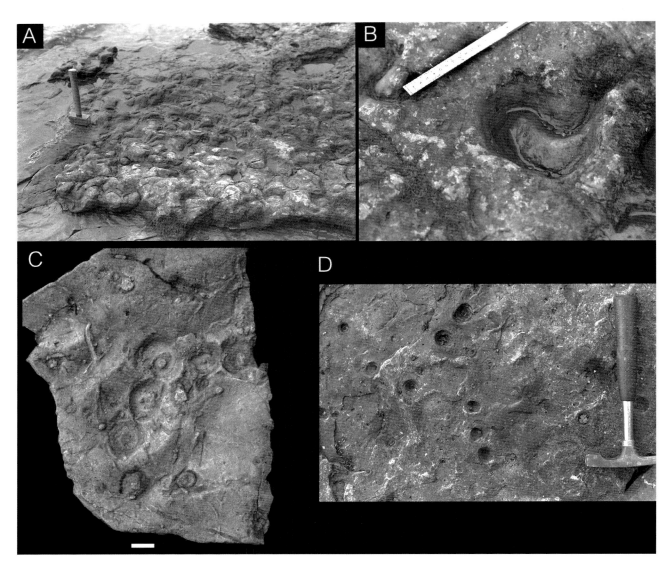

Fig. 6. □A. Field photograph of medium-grained sandstone beds with abundant weathered, curved *Rhizocorallium jenense*. Upper part of interval C at Hällekis. Length of hammer-shaft 22 cm. □B. Close-up of weathered specimen of curved *Rhizocorallium jenense* from same outcrop as in A. □C. Field photograph of slab from Trolmen harbour with *Monocraterion* isp. Probably from the top of interval A, or base of interval B. □D. Field photograph of *Monocraterion* isp., about 400 m north of Trolmen harbour. Top of interval A.

grains (Fig. 5B). X-ray diffraction confirms the presence of siderite. Owing to weathering, the clasts are often reddish brown. At Lugnås they are most common in the upper part of Interval A, where the sediment is often developed as coarse sandstone. The only body fossil reported from the conglomerate is *Torellella laevigata* known from Lugnås (Linnarsson 1871; Holm 1893; Hadding 1927, p. 56) and Kinnekulle (Lindström 1977). Horizontal and flatly U-shaped spreite burrows are numerous on the base of the upper sandy part, including poorly preserved *Rhizocorallium*. Also, in the lower beds, there are arthropod trace fossils. Incomplete shrinkage cracks are observed on bases of sandy beds (Fig. 5C). About 300 m north of Trolmen harbour there are large-scale asymmetric ripples with a wavelength of about 60 cm, a ripple index of 12, and striking at NE–SW. They consist of quartz-pebble-bearing, coarse-grained sandstone. In the troughs there are smaller, poorly visible interference ripples. Patches of these beds are again visible 600 m north of Trolmen harbour. These beds have *Monocraterion* (Fig. 6D) and *Diplocraterion*. North of Råbäck, Högbom & Ahlström (1925, p. 79, Fig. 15) described and illustrated large ripples with gently lunate and discontinuous crests immediately above the basal surface.

Interpretation. – Trace fossils supposedly made by trilobites indicate marine conditions for the basal part of the Mickwitzia sandstone. The basal conglomerate has long been interpreted as transgressive over the denuded basement. The clasts in the conglomerate can largely be assigned to the local basement, including some from pegmatite veins (Högbom & Ahlström 1924; Hadding 1927). Among possible intraformational components are clastic rocks that may be from an earlier stage of the transgression, including the sideritic mud clasts mentioned above. It has been suggested that the sideritic elements represent remnants of continental deposits (Westergård 1931b), but siderite also forms in marine environments, and sideritic clasts are a common constituent of intraformational conglomerates.

The small thickness of the conglomerate and its interlayering with pure sandstones and mud layers indicate that it was formed under conditions of slow sedimentation and may be a transgressive lag deposit. The agent of winnowing may have been strong waves or currents formed by storms (cf. Anderton 1976; Levell 1980). Asymmetrical large-scale ripples near the base may be more or less shoreline-parallel sand dunes, formed in the offshore to foreshore. Anderton (1976) interpreted megaripples in the Jura Quartzite, England, to form from small migrating sand dunes, which may also be the case here.

Interval B, 0.3–3.8 m

Description. – Thin-bedded sand- and siltstones with thin clayey layers. Sand- and siltstone beds mostly less than 7 cm thick with sharp bases and with horizontal or near-horizontal lamination, at the top turning into wave-ripple lamination. Tops of beds mostly with ripples of interference type, though straight-crested ripples also occur. Beds typically have an upward fining with coarse-to medium-grained sandstone in the lower part, towards the top turning to fine sandstone and siltstone. Laminae with coarse to medium sand decrease in thickness towards the top of the bed, where finer-grained laminae dominate. Horizons rich in mica flakes occur, usually associated with fine-grained sediment. The beds may have undulating bases with laminae following the contour of the lows or meeting the base at a low angle. This gently convex-down lamination is generally cut on both sides of ripple highs by bundles of laminae following the upper surface. Such bundles may also continue into the trough and drape adjacent cosets. Often there is a strongly developed superposition. This characteristic bedding type may be classified as lenticular bedding of linsen-type (deRaaf *et al.* 1977).

At Hjälmsäter there is a 10 cm thick bed at 1.0 m above basement, and immediately above this there is an approximately 30 cm thick sandstone bed that forms the base at the section at Hjälmsäter. The top of these beds is gently undulating. A portion of such a bed was illustrated by Högbom & Ahlström (1924, Fig. 5). The upper surface shows interference-type ripples. In the Mossen core, thick beds at 0.7 m have coarse material at the base, with 2 cm long quartz pebbles. At about 0.4 and 0.7 m are decimetre-thick levels with thin siltstone lenses alternating with thin clayey layers. Clayey layers are mostly below 2 cm, but some are thicker, including 6 and 9 cm thick beds at 0.5 and 0.8 m, respectively. According to Lindström & Vortisch (1978), these are silty kaolinite- and illite-containing clays. Several levels show disrupted bedding due to soft-sediment deformation and burrowing. The uppermost part of the interval has beds, a few centimetres to some decimetre thick, with slanting bedding-planes and amalgamation.

Interval B is rich in well-preserved trace fossils, including *Cruziana rusoformis*, *Cruziana tenella*, *Diplocraterion parallelum*, *Gyrolithes polonicus*, *Palaeophycus imbricatus*, *Olenichnus isp*, *Rosselia socialis*, *Rusophycus dispar*, *Rusophycus jenningsi*, *Trichophycus* isp. and *Zoophycos* (*Rhizocorallium*) isp. The degree of bioturbation is low to moderate (Ichnofabric indices 2–3; see Droser & Bottjer 1986). Physical sedimentary structures include *Aristophycus*-like forms, flute casts, load casts, obstacle marks, pot- and gutter casts, spill-over ripples, incomplete shrinkage cracks, Kinneyian marks and tool marks, including 'Eophyton' (Figs. 5, 7–11). Body fossils consist of *Mickwitzia monilifera*, *Volborthella tenuis*, *Mobergella* sp, and *Torellella laevigata*. *Volborthella* may occur in great numbers and also occurs within the beds to form coquina-like beds. Concentrations are especially high in pot and gutter casts. In pot-casts, specimens are found also in an upright position with the narrower end directed down. This orientation does not necessarily indicate life position but may be the result of settling from suspension. Transport is suggested also by the abraded appearance and breakage of the pointed end of *Volborthella* and by mud-filled *Mickwitzia* shells embedded in otherwise sandy sediments. *Volborthella*-rich seams are also found within the beds occurring along erosional surfaces (Fig. 15A). *Mobergella* is known only from Kinnekulle, where it is found at a level 0.3–1 m above the basement, in layers especially rich in problematic structures, including *Spatangopsis* and *Protolyellia* (Figs. 12–13). These surfaces are found on trough-like sand bodies with flat to undulatory basal outline and sharp and steep borders. In addition to numerous types of problematica, there may occur tool marks and irregular surfaces with coarse-grained material, probably deformed flute casts or load casts (Fig. 14). The beds are often built up of thin layers that exhibit strong amalgamation and highly wavy contacts. Individual layers are rarely thicker than 5–8 mm, are typically separated by some muddy material, and consist of fine-grained sand. Some levels consist of alternating silt and mud with silt laminae less than 1 mm. There are also thin seams of coarser-grained sediment within the beds.

Fig. 7. Sedimentary structures from the Mickwitzia sandstone. □A. Slab with three sandstone beds exhibiting erosional contacts. Lower bed consists of fine sandstone with nearly horizontal lamination. On the left side are remnants of a fine-grained sandstone with a load-casted base. This is cut by the top bed, which consists of medium-grained sandstone, with occasional larger quartz fragments, and mud clasts. The top of this bed has ripples with a wavelength of about 15 cm and trace fossils including *Rhizocorallium jenense*. The upper bed truncates a *Diplocraterion parallelum*. Stora Stolan, Billingen. Scale bar 10 mm. RM X3291. □B. Lower surface of slab in A, with tool marks and flute casts, the latter showing signs of deformation. Scale bar 10 mm. □C, D. Lower (C) and side (D) view of a pot cast that at its center has a shell of *Mickwitzia monilifera*. Sides of the pot cast have imbricated ridges. Scale bar 10 mm. Hjälmsäter. Level B (0.4–1 m above basement). RM X3232.

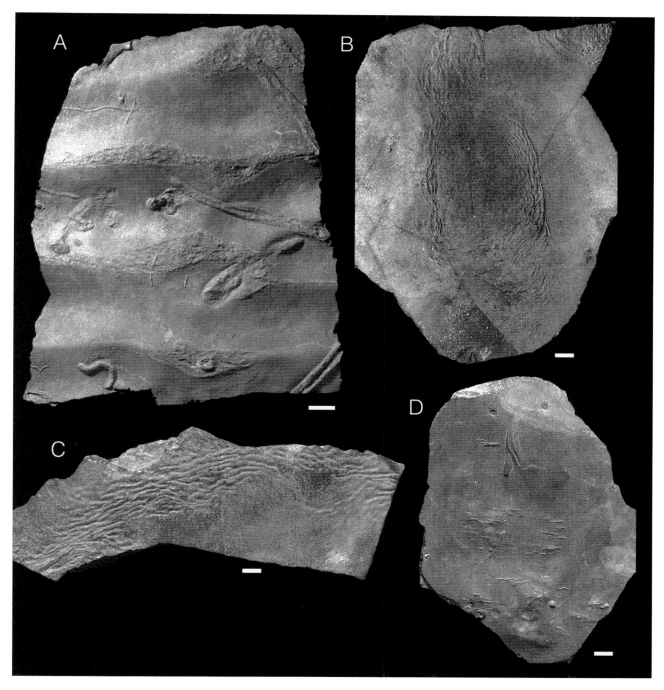

Fig. 8. Sedimentary structures from the Mickwitzia sandstone. □A. Kinneyia marks capping the top of rippled sandstone bed. Also seen are trace fossils, including *Palaeophycus imbricatus*. Scale bar 10 mm. Specimen in the collection of Jan Johansson, Sköllersta. □B. Base of a laminated fine sandstone–siltstone with deformational structures. Scale bar 10 mm. Hällekis, probably interval D. RM X3292. 20. □C. Wrinkled lower surface on horizontally laminated, fine sandstone–siltstone. Scale bar 10 mm. Hällekis, interval D. RM 3293. □D. Base of fine sandstone–siltstone with filled tensional fractures. Near top of slab is a *Palaeophycus imbricatus* associated with tension fractures caused by the burrowing animal. The actual, smooth-walled, burrow is seen as a cylindrical tube. Scale bar 10 mm Lugnås. SGU 8578.

Interpretation. – Beds with a sharp erosional, flat to gently undulating base, horizontal to low-angle lamination, graded bedding, and interference type wave ripples, most likely represent sediment deposited by storms. The presence of wave ripples suggests deposition above storm-wave base, with the ripples forming at the waning phase of the storm (Aigner & Reineck 1982, Aigner 1985; Brenchley et al. 1986). Polygonal and other wave ripples were reported to be common in distal parts of storm-deposited layers formed on a wide Ordovician shelf in Spain

(Brenchley et al. 1986). Hummocky cross-stratification has not been identified, but some thick beds with gently undulating surfaces are possible candidates.

Other features indicative of a high-energy hydrodynamic setting for the Mickwitzia sandstone, and possibly indicative of storms, are pot and gutter casts, Kinneyian marks and spill-over ripples (Seilacher 1982). According to Aigner & Futterer (1978), pot-casts arise by erosion around an obstacle, which may develop into a circular hole and expand downstream into a gutter. In the basal part of the pot cast, an elevation indicates the position of the obstacle, which may still be in place. In the Mickwitzia sandstone, shells of the brachiopod Mickwitzia have been found at the base of pot casts and may have served as an obstacle (Fig. 7C, D). However, an obstacle is probably not necessary for the formation of pot cast; they may form also by spiral vortices (see Myrow 1992). Small-scale channels with densely tool-marked surfaces, probably corresponding to gutter casts of Aigner & Futterer (1978), occur together with the pot casts, and conjoined examples of the two have been found in the Mickwitzia sandstone. Seilacher (1982) considered the absence of these structures in modern muddy bottoms to indicate formation under sand-saturated flow.

The setting for this interval is interpreted as shallow shelf with domination in time of sedimentation of mud, and intervals of influx of coarser-grained sediments. The lower part of interval B is characterized by frequent gutter casts in a generally fine-grained sediment. Lags and seams of coarse material and frequent amalgamation indicate reworking and winnowing. Trace fossils conform to the *Cruziana* ichnofacies, with assorted dwelling and feeding burrows.

Interval C, 3.8–7.8 m

Description. – Medium-grained sandstone. Most beds in this interval consist of medium-grained, friable, sandstone with beds 5–10 cm thick, some up to 30 cm, with common diffuse bed boundaries because of a high degree of bioturbation. The dominating material is well-sorted

and well-rounded quartz. Haematitic staining around quartz grains is common, and pyrite also occurs. At several levels there are flat muddy intraclasts, up to several centimetres in length, with the highest concentration in the lower parts of beds. Carbonate cement is more common than in interval B. Bioturbation is higher than in interval B (ii 3–4), occasionally resulting in a knolly or massive appearance. Intensively bioturbated beds (ii4–5) are particularly conspicuous in the lower part of this interval. Trace fossils include generally poorly preserved but locally abundant *Rhizocorallium jenense*, *Halopoa imbricata* and, more rarely, *Rusophycus dispar* and *Syringomorpha nilssoni*. Of body fossils, only fragmentary *Mickwitzia* are found. In the Hjälmsäter profile there are several harder beds at the top of the sequence with erosive bases cutting into the underlying beds. Phosphatized mud clasts were reported by Möller (1987) at 6.6 m above the basement. At Hällekis there are beds about 6 m above the basement with abundant *Rhizocorallium jenense* and phosphatic mud clasts (Fig. 6A, B).

Interpretation. – In interval C the proportion of sandy to clayey layers increases. This is interpreted as reflecting shallowing, possibly entering or approaching a sand flat affected by storm and possibly by tidal currents. Erosion of the seafloor is indicated by ripped-up mud clasts. The intense bioturbation may be responsible for the friable character of the sediment, as muddy material was incorporated into the sand, which may have hindered or made cementation more difficult.

Interval D, 7.8–9.8 m

Description. – Mixed layer. This interval has a wide range in bed thicknesses and grain sizes and also is the most difficult to compare between sections. Several levels have thin-bedded sand and siltstone beds, occasionally of lenticular/flaser-bedding type (Figs. 14A, 15B). These beds have abundant trace fossils including *Rusophycus dispar*, *Cruziana rusoformis*, *Teichichnus* s.l., and *Cruziana tenella*. Scholz (1977), and Möller (1987) listed an occurrence of *Spatangopsis* within this interval. At Hällekis, planar beds with horizontal lamination have bases possessing a pattern of parallel to arcuate wrinkles (Fig. 8C). In the Hjälmsäter profile there are thicker sandstone beds at 7.0–7.5 m (Fig. 16B). The uppermost of these beds at Hjälmsäter has large-scale ripples striking NW 295°–336° with a wavelength of 25–30 cm and an index of about 8; current direction is from the west. In the upper part of interval D there are several rippled layers with coarse material up to 5 mm in diametre and some thicker layers with a loose conglomeratic composition (Figs. 16A, 17A). At Hjälmsäter there is a conglomeratic bed at 8.4 m. At Hällekis these beds include pebbles of quartz. In the Hjälmsäter

Fig. 9. Shrinkage cracks from the Mickwitzia sandstone. □A. Several generations of filled shrinkage cracks. The straight structure with a horizontal orientation is a *Palaeophycus imbricatus* which predates the cracks. Scale bar 10 mm. Lugnås. SGU 8579. □B. A folded sheet-like filled crack on the top of a rippled bed. Scale bar 10 mm. Stora Rud, Lugnås. RM X3294. □C. Extensive shrinkage cracks on base of bed of fine sandstone. Lower part of Level E. Hjälmsäter. RM X3295. □D. Shrinkage cracks with sharp V-shaped profile, developed into rough T-shape or cross-patterns. These forms could be crystal pseudomorphs, Scale bar 10 mm. Level B (0.4–1 m above basement) Hjälmsäter. RM X3201. □E. Shrinkage cracks formed in connection with deformation of a flute cast. Scale bar 10 mm. Lugnås. SGU 8580.

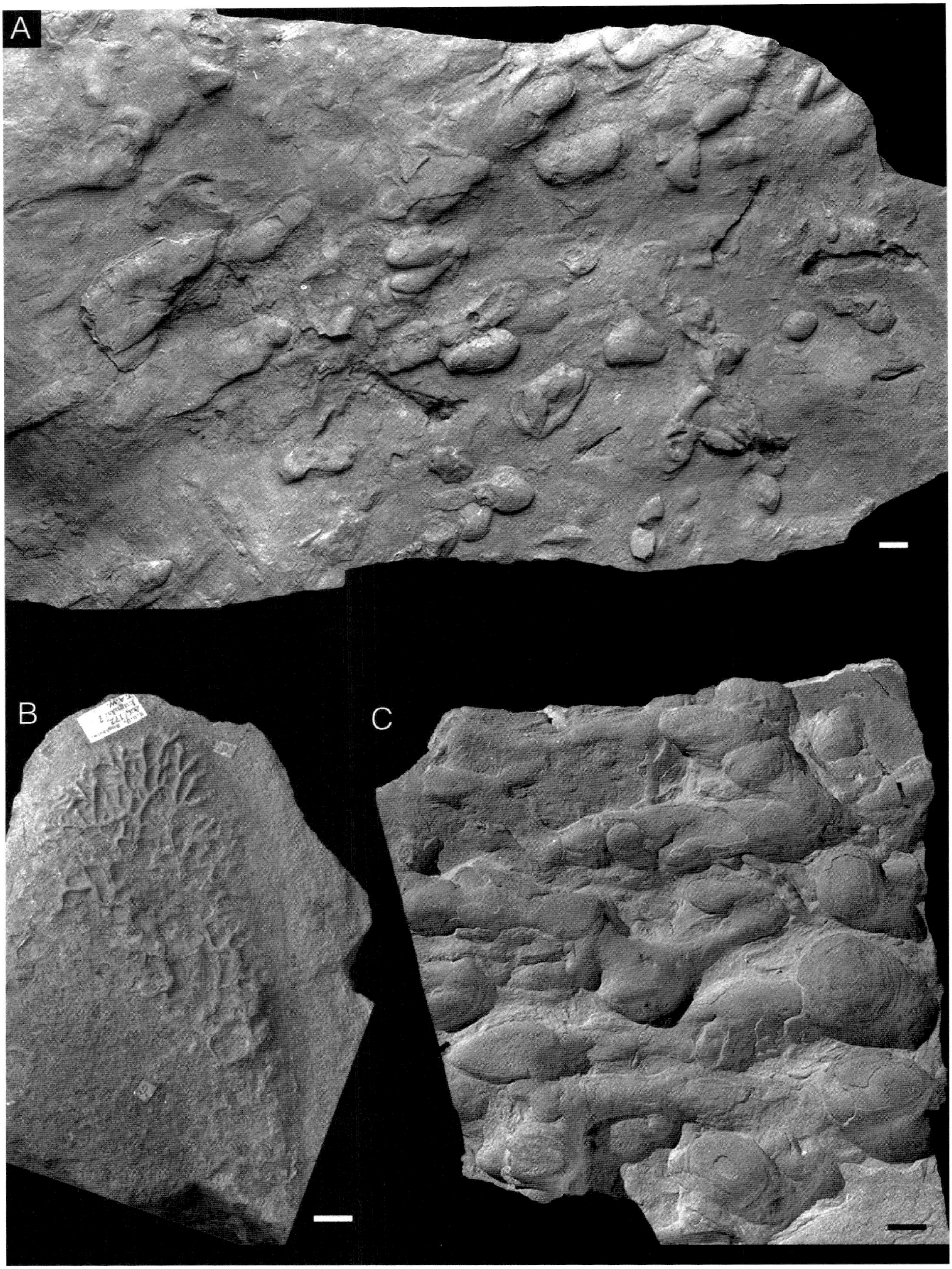
A
B
C

profile, shrinkage crack casts are abundant at about 8.7–9.0 m, including nearly polygonal patterns (Fig. 9C). Such filled cracks were also observed at Trolmen, where they occur at 9.6–9.8 m, and in the Mossen core at 9.4 m. Several of the sandstone beds in the upper part are spotty, due to oxidized iron minerals.

Interpretation. – This interval is noticeable for containing coarse-grained rippled beds, with beds in the upper part of the interval including sparse conglomeratic development. The coarse-grained rippled beds are similar to, though generally smaller than, wave-formed coarse-grained ripples, discussed and reviewed by Leckie (1988). These coarse-grained beds occur within a bedding of the type seen in interval B, with thin sand–silt and clay beds, rich in trace fossils including large sand-filled burrows of *Teichichnus/Trichophycus* type (Fig. 17B). The reported occurrence of a *Spatangopsis* would further accentuate similarities with interval B, though this finding needs to be substantiated. Filled shrinkage cracks are developed in greater numbers and with more nearly polygonal shape than anywhere else in the Mickwitzia sandstone. However, these are also probably of subaqueous origin, formed as a result of soft-sediment deformation (see below).

Interval E, 9.8 m to about 11 m

Description. – Thick-bedded sandstone. Sandstone beds of highly variable vertical thickness. Beds up to 20 cm or more in thickness, with abrupt, laterally discontinuous divisions into thinner beds with clayey partings (Fig. 17A,

C). The lowermost part of this interval consists of coarse-grained sandstone followed by clayey beds with abundant sandstone-filled shrinkage cracks (Fig. 5A). Of trace fossils, *Skolithos* is abundant (ii 3–4; see Droser & Bottjer 1989), while *Diplocraterion* is less common. A pronounced parting is observed in the Mossen core at 10.5 m, marked by a distinctly uneven surface, also visible at Hällekis and Trolmen. Sediment immediately above this surface is coarse-grained and again more coarse-grained at 10.8 m. Spots rich in carbonate cement may occur and upon weathering result in numerous centimetre-sized pockets. These beds are cross-bedded with truncating sets of laminae (Fig. 17C) and no consistent dip seen in outcrops at Hällekis. Clay intraclasts occur. A greenish tint to the rock in the upper part of the interval has been taken to be from glauconite (Ahlberg *et al.* 1986) Above this level, from about 11 m, the sediment turns to more consistently fine-grained sandstone, which continues through the Lingulid sandstone. Level E forms a thick package overlying thinner-bedded parts and is easily recognized in the field.

Interpretation. – This interval probably represents shoreface deposition. Erosion is seen from incorporated mudclasts and possibly from the uneven surface at about 10.5 m. The dominant bedding type in interval E is similar to steep hummocky cross-stratification with erosivly removed tops (e.g., Brenchley 1985, Fig. 5), to which it is tentatively assigned. The common occurrence of *Skolithos* may represent rapid colonization by opportunistic animals (cf. Droser 1991).

Selected sedimentary structures

The following section presents and discusses a few sedimentary structures of particular interest or potential importance for the interpretation of the sedimentary environment.

Tool marks. – Tool marks are ubiquitous on the bases of beds, especially in interval B. There is a wide range of forms, the morphologic variation of which can be attributed to the angle of impact between the tool and the clay surface. Especially conspicuous are elongated flat bands

---

Fig. 10. Sedimentary structures from the Mickwitzia sandstone. Scale bars 10 mm. □A. Base of bed with deformed flute casts. Slits cutting through the base are *Diplocraterion parallelum.* Stora Stolan, Billingen. RM 3296. □B '*Aristophycus*'-like structure on base? of bed which in the upper? part shows extensive soft-sediment deformation. Lugnås. SGU 8581. □C. Flute marks on base of fine-grained sandstone. Hjälmsäter. Interval B. RM X3352.

up to 6 cm wide, with longitudinally striated surfaces, reported in the literature as '*Eophyton linnaeanum*' Torell, 1868, and '*Eophyton torelli*' Linnarsson, 1869 (Figs. 11, 13A, 46B). These structures were interpreted as plant remains by Torell (1868, 1870) and Linnarsson (1869a, 1871). Nathorst (1874) considered this unlikely and invoked current-transported algae to explain shallow forms of '*Eophyton*', whereas he suggested several possible modes of origin for deep forms, including dragging of stones or activities by crustaceans, sand worms or echinoderms. Seilacher (1992a, 1994a) claimed that the continuous bands of '*Eophyton*' are unique among sandy tempestites and suggested that '*Eophyton*' in the Mickwitzia sandstone were produced by sand skeletons transported by storm waves (Seilacher 1992a, p. 612). Tool marks similar to those in the Mickwitzia sandstone have been illustrated by Dżułyński & Sanders (1962, Pl. 11B, 20B) and Reif (1982, Fig. 4a). The notion that '*Eophyton*' tool

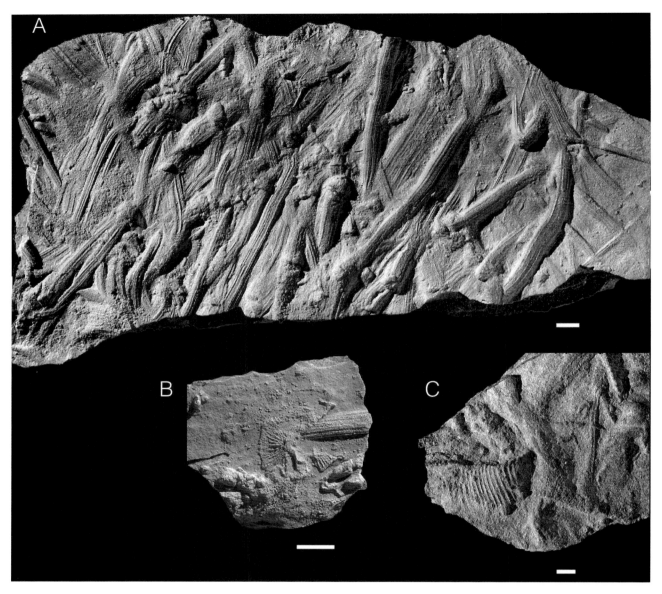

Fig. 11. Sedimentary structures from the Mickwitzia sandstone. Scale bars 10 mm. □A. Tool marks of '*Eophyton linnaeanum*'-type. Lugnås. RM X147. □B. Base of bed with '*Eophyton*'-type tool mark, in close proximity with fan-shaped ridges that possibly represent frondescent casts. SGU 8582. □C. Tool marks. To the left is a prod mark with parallel ridges. Hällekis, Level D. RM X3297.

marks could be used as time markers in the Baltic area (Seilacher 1992a, p. 612), is optimistic; the similarities in sedimentary structures (see below) and trace fossils between the Mickwitzia sandstone and the Lükati Sandstone in Estonia depends on similar conditions of sedimentation. No tool has been found directly connected to an 'Eophyton' in the Mickwitzia sandstone, and their nature cannot be decided. Possible objects, however, are intraclasts. 'Eophyton'-type tool marks change from broad bands to narrow erect ridges, implying a nonequilateral object. Intraclasts in the Mickwitzia sandstone enclose grains of coarser material, up to medium sand. The deli-

cate striation seen on 'Eophyton' may have been formed by these grains, when intraclasts were transported within sediment-laden storm currents.

Another common tool mark consists of delicate ridges arranged in parallel sets (Figs. 7B, 34A; Nathorst 1881a, Pl. 6:3; Hadding 1929, Fig. 22). This type is easily confused with *Monomorphichnus*, and both may be formed by an arthropod caught by currents.

Kinneyian marks. – Martinsson (1965) introduced the term 'Kinneyian ripples' for small ripple-like structures similar to supposed algal structures described by Walcott

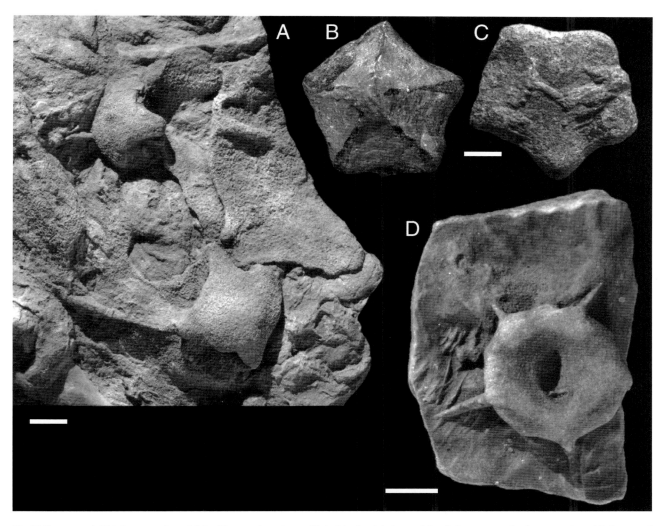

Fig. 12. Spatangopsis. □A. Lower surface of slab with several specimens. Note thin sheets in lower right of picture occurring in connection with the central sandy body. Scale bar 20 mm. RM X3237. □B, C. Upper and lower surface of a *Spatangopsis* which at its base preserves a shallow *Rusophycus.* Scale bars 10 mm. Lugnås. SGU 8619. □D *Spatangopsis* protruding from base of slab. Scale bar 20 mm. Hjälmsäter. RM X3350.

(1914). In the following, the term Kinneyian marks is used to avoid the genetic implications in the term Kinneyian ripples. Singh & Wunderlich (1978) referred Kinneyian marks to wrinkle marks (= Runzelmarken), while Walcott's 'Kinneyia' was referred to millimetre-sized ripples; the two differ by an irregular pattern of crests in the former and more parallel-running crests in the latter. Based on comparison with similar structures formed experimentally and observed on modern sediments, these structures have been considered indicators of (quasi-)subaerial exposure, formed by strong winds blowing over a cohesive sediment surface covered by a thin film of water (Reineck 1969) or by raindrop impressions reworked by wave swash (Klein 1977). Allen (1985) described wrinkle marks formed in intertidal settings of the Severn Estuary and Bristol Channel. These were inter-preted to form as post-depositional load structures caused by pore-water seepage in exposed fine layers of silt and mud.

Kinneyian marks in the Mickwitzia sandstone are found on tops of rippled surfaces, where they occur on planar surfaces sharply truncating the ripple (Fig. 8A). Similar Triassic forms from Iraq were illustrated by Seilacher (1982), who considered oversteepened profiles to indicate formation within the sediment, as load-induced phenomena related to event sedimentation. He suggested that the ripples formed after the surface was mud covered, possibly related to differential dewatering. A slab with Kinneyian marks from the Mickwitzia sandstone was figured by Wiman (1943, Fig. 4). On this slab, the Kinneyan marks are found together with oval, shallow troughs possessing concentric lamination. Seen in profile,

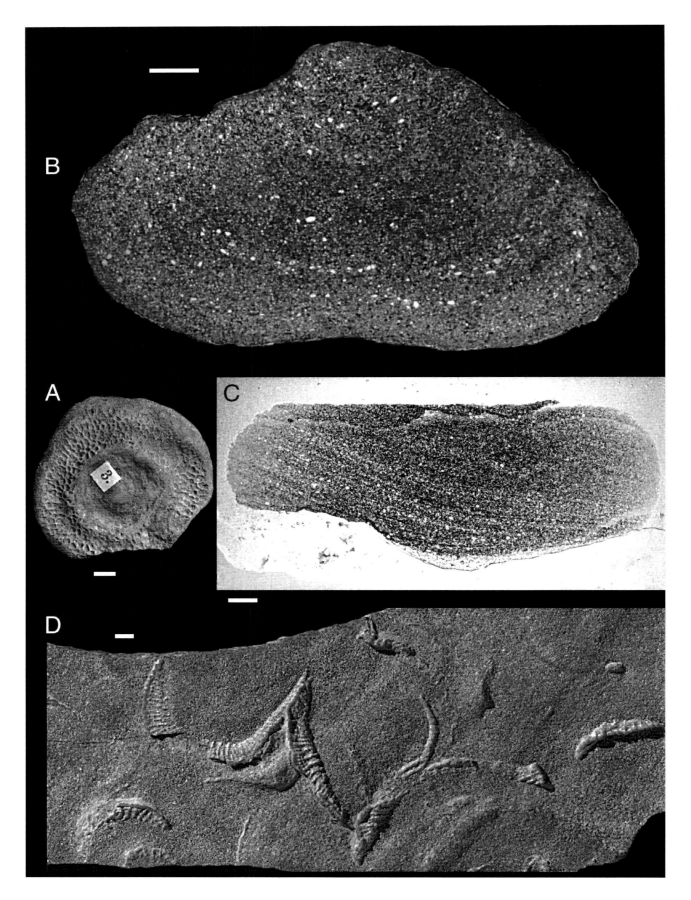

this lamination continues within the sediment as upward-fining, gently concave-upward bowls. Under the Kinney-ian marks there is also inclined lamination. Ridges crowd near the ovals and also show some deflection. These ovals appear to be the troughs of planed-off ripples. Association with inclined laminated beds suggests that Kinneyian marks form when a laminated rippled surface is eroded as a result of the material in the laminated sediment being differently susceptible to erosion and deformation.

There are probably several different modes of origin for wrinkle marks and similar structures. Forms generated under conditions of quasi-emergence as well as those resulting from load phenomena under intertidal conditions have wrinkles formed also in ripple troughs. On the other hand, those from the Mickwitzia sandstone and those illustrated by Seilacher (1982, Fig. 4a) occur on ripple tops and are interpreted to have formed intrastratally. Examples of Kinneyian marks reported in sediments probably formed under influence of storms are the Lower Cambrian Lükati Formation of Estonia (personal observation, 1992), the Middle Cambrian Paradoxissimus Silt-stone of Öland, Sweden (Martinsson 1965), *Diplocraterion yoyo* facies of the Late Devonian Baggy Beds, north Devon, England (Goldring 1971), the upper Unit of a sequence of Upper Devonian Catskill clastic wedge, Pennsylvania, USA (Bridge & Droser 1985). The strongly planed-off appearance suggests that these Kinneyian marks were formed by currents eroding fine-grained cohesive sediments.

Shrinkage cracks. – Filled shrinkage cracks occur throughout the Mickwitzia sandstone (Figs. 5A, C; 9), at most levels as isolated, thin, sheet-like to lenticular structures (Fig. 9B). Where cross-cutting relations with trace fossils have been observed, the shrinkage cracks postdate the trace fossils. Incomplete cracks with bird-foot patterns are found in the basal conglomeratic part (Fig. 5C). More extensively developed crack patterns are found in some beds of the upper part of the Mickwitzia sandstone, especially in the top of interval D, where they may form irregular net-work patterns (Fig. 9C).

The interpretation of shrinkage cracks is problematic as, in addition to formation due to desiccation by subaerial exposure, there have been numerous reports of cracks formed subaqueously. The latter interpretation has usually been applied to spindle-shaped or fusiform, isolated to incompletely formed forms, though polygonal or near

polygonal patterns have also been assigned (e.g., Ricci-Lucchi 1970, Pl. 133). A review of the subject was given by Plummer & Gostin (1981). They stressed the importance of soft-sediment deformation in the origin of subaqueous shrinkage cracks. Structures of undoubted subaqueous origin, formed in connection with sediment deformation, are found also in the Mickwitzia sandstone (Fig. 9E). Such forms may also be termed sedimentary or clastic dikes (e.g., Allen 1982; Reineck & Singh 1980). Related to these are probably also branching to dendritic ridges typically preserved on the top of beds (Figs. 5E, 10B; see also Hadding 1929, Fig. 32). According to Hadding (1929, p. 52) these are the forms assigned the name '*Archaeorrhiza*' by Torell, 1870, and some more intricate forms are similar to '*Aristophycus*' (Fig. 10B; see Häntzschel 1975, p. 169). These may have formed as sedimentary intrusions within the sediment owing to instability. Seilacher (1982) suggested that '*Aristophycus*' structures was formed by pore water leaving sand preferentially through burrows and ripple crests and carving out the intricate pattern. The lack of size differentiation in the Mickwitzia sandstone specimens makes a similar explanation questionable.

Astin & Rogers (1991) questioned the subaqueous interpretation of lenticular, straight to curving cracks from the Devonian Orcadian Basin in Scotland, and they also questioned the reports of subaqueous cracks in the literature. They reinterpreted such cracks to form either as incomplete desiccation cracks or cracks initiated by inhomogeneities caused by gypsum crystals.

Polygonal development is generally regarded as indicating subaerial formation. However, a wide variety of crack patterns may be formed, depending on the degree of desiccation and properties of the sediment (e.g., Allen 1987). Thin, isolated forms, common in interval B of the Mickwitzia sandstone, are interpreted as intrastratally formed due to dewatering, in most cases probably as a consequence of soft-sediment deformation at an early stage. Similar forms from Cambrian deposits of the Holy Cross Mountains were interpreted to have formed as a result of stress in the sediment (Radwański & Roniewicz 1960, Pl. 30:2). The strongest case for crystal-initiated cracks are specimens such as that on Fig. 9D. These are roughly T-shaped, nail-shaped, or pointed cross-shaped, with a V-shaped vertical profile. The shape is suggestive of gypsum crystals. What seems to be similar structures were illustrated by Dżułyński & Żak (1960) from the Cambrian of the Holy Cross Area. These were interpreted as formed by burrowing animals, though this seems highly questionable. It is uncertain if any of these cracks were formed or initiated by crystals, but Lindström (1977, 1984) has interpreted structures from the basal layers of the Mickwitzia sandstone as evaporite or calcite pseudomorphs.

The most likely candidates for subaerial formation are the shrinkage cracks at the top of interval D. The levels in which the most highly developed cracks occur are in close

Fig. 13. Protolyellia. □A. Upper surface of specimen with well-developed net- to honeycomb-shaped pattern. Lugnås. Scale bar 5 mm. SGU 8613. □B. Micrograph of thin section of *Protolyellia* with crude convex-up lamination. Lugnås. SGU 8620. □C. Micrograph of thin-section of sharp-based bed from interval B. Compare with B. Hjälmsäter. Scale bar 2 mm. SGU 8621. □D. Problematicum on base of micaceous, about 5 mm thick bed of medium-grained sandstone. Scale bar 5 mm. Specimen in collection of Jan Johansson, Sköllersta.

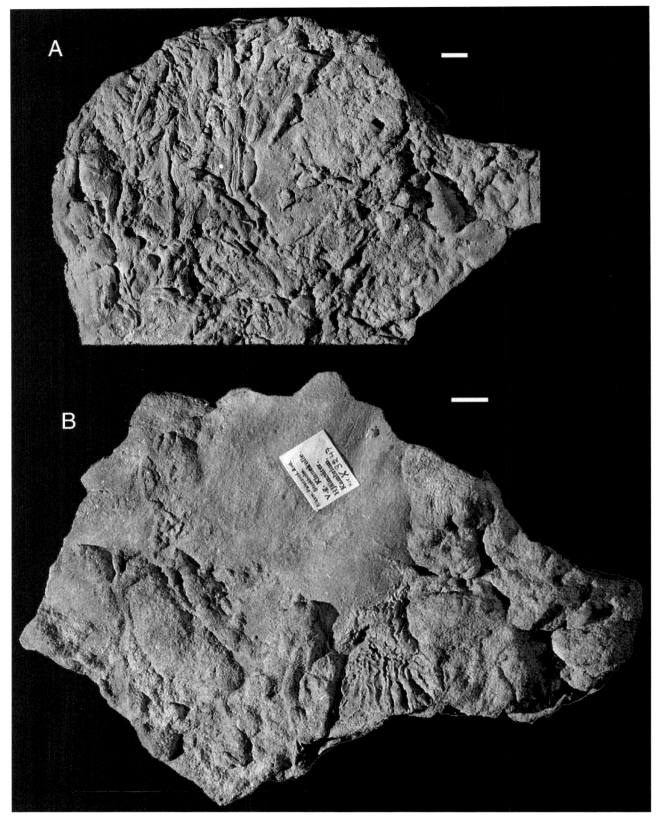

Fig. 14. Soles of beds from the lower part of interval B, 0.4–1 m above basement, at Hjälmsäter. Scale bars 10 mm. □A. Bed with tool marks and sandy sheets. Some of the latter with punctate structures. See also Fig. 14. RM X3205. □B. Base of bed with irregular surface caused by fluting and/or load casting. Note sharp ridges. RM X3247.

Fig. 15. Thin-sections of beds from the Mickwitzia sandstone. Scale bars 2 mm. □A. Bed consisting of two amalgamated beds, the upper with coarse base and scattered *Volborthella tenuis*. Contact between the two beds irregular due to minor slumping. Near the base of the lower bed is a section through a poorly preserved *Mickwitzia*. Trolmen, interval B. PMU 993 □B. Sandstone with graded bedding from Level D at Hällekis. Vertical structure probably a burrow. PMU 994.

association with sediments interpreted as the shallowest-water sediment in the Mickwitzia sandstone. A possible cause of periodic exposure in shallow seas was given by Ginsburg (in Dott & Byers 1981), who suggested that strong persistent winds could have caused tilting of the water mass, resulting in subaerial exposure. This could have caused desiccation shrinkage cracks, and the sloshing of water could also have ripped up mud, now found as intraclasts (Ginsburg in Dott & Byers 1981; Bose & Chaudhuri 1990). This mechanism could account for polygonal crack fills from the late Cambrian Jordan Formation, Wisconsin, found in layers indicating conditions in which desiccation cracks are unexpected (Dott & Burgeois 1982). This process may also have produced the shrinkage cracks in the Mickwitzia sandstone.

'Sand corals'. – The Mickwitzia sandstone contains a number of structures for which the interpretation is problematic, even to the point that it is uncertain whether the structures are of organic origin or not. They lack any signs of biomineralization but have shapes that are unusual for inorganic sedimentary structures or trace fossils. A detailed investigation of these structures will be presented elsewhere. Here I will briefly discuss features relevant for the interpretation of the most widely known of these forms, *Spatangopsis* Torell, 1870, and *Protolyellia* Torell, 1870, which have been considered core-members of an extinct group of Cnidarians (Seilacher 1994). Both consist of siliciclastic material, and typically they are found as three-dimensional bodies within muddy beds of the heterolithic interval B. *Spatangopsis* consist largely of medium-grained sandstone; in *Protolyellia* the sediment often occurs in crude cup-shaped laminae distinguished by grain-size variations. *Spatangopsis* is typically pyrami-

dal to star-shaped, with a broad base and upward taper (Fig. 12). Four to six ridges radiate from the highest point, delineating flat, convex or concave h. *Protolyellia* has a convex lower surface and an upper surface that is flat, gently convex or raised towards the center (Fig. 13). The upper surface has concentric ornamentation and/or radiating to honey-comb shaped patterns formed by narrow ridges (Fig. 13A). Seilacher (e.g., 1983a, 1992a, 1994a) has developed the idea of Psammocorallia, an extinct group of anthozoans forming a stabilizing sand skeleton by phagocytizing sand grains. The sand skeleton also served to form endodermal septa.

Since the sand bodies were supposedly held together only with organic material and would disintegrate rapidly after the organism's death, embedding within muddy sediments was considered a necessary condition for their preservation (Seilacher 1992a). Seilacher (1992a) related the preservation to obrution with smothering mudfall following storms. However, as seen from slickensides, these bodies were well-embedded within the mud, and on some specimens of *Spatangopsis* the ridges adhere to the sole of the overlying bed. In these, as well as specimens still embedded in mud, the ridges ('septa') continue beyond the main body, often exhibiting varying degrees of folding and slumping. In specimens that adhere to sole surfaces the ridges meet this surface at about right angle. This shows that the ridges were not enclosed by an organic envelop but were introduced from above. Another feature that makes the sand coral interpretation unlikely are specimens with a basal wrapping possessing structures formed at the base of a bed, including *Rusophycus* trace fossils.

There are thus several indications that *Protolyellia* and *Spatangopsis* formed by sediment that foundered into

Fig. 16. Bedding from upper part of the Mickwitzia sandstone. □A. Section at Hjälmsäter showing beds of interval D and E. The top of the bed in the lower part of the figure is about 8.5 m above basement. This bed, about 20 cm thick, is locally conglomeratic. □B. Section at Hjälmsäter showing interval C and the upper part of interval B. Base of the picture shows beds about 2 m above basement; top shows beds at about 8 m. □C. Erosive contact between beds from interval D at Hjälmsäter. Hammer placed in ripple trough about 7.5 m above basement. Length of hammer shaft 22 cm.

mud similar to the formation of pseudonodules (see Richter 1971; Allen 1982). This also explains the cup-shaped laminae in *Protolyellia* as originally horizontally laminated beds (Fig. 13B, C).

Another unusual structure in the Mickwitzia sandstone consists of crescent-shaped sand sheets possessing distinct ridges, often developed as a string of beads (Fig. 13D). Seilacher (1994a) suggested that these are sand sponges. The sheets meet the base of the bed at a high angle. This orientation is difficult to reconcile with a smothered 'sand skeleton' but is consistent with sand injection.

Sedimentary environment – discussion

The Mickwitzia sandstone is one of several late Protero-zoic – early Palaeozoic sandstone units lying on a low-relief cratonic basement. These sandstones have small vertical extent but may be continuous for large areas (e.g., Dott & Byers 1981). The number of studied exposures of the Mickwitzia sandstone is small but their general similarity in order of lithologic changes is consistent with formation in a wide shallow sea. Högbom & Ahlström (1924) pondered that most of the material in the Mickwit-

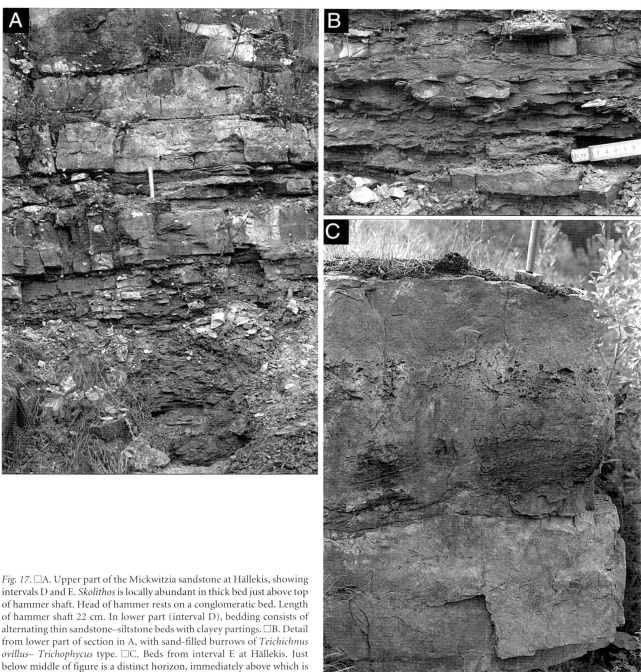

Fig. 17. □A. Upper part of the Mickwitzia sandstone at Hällekis, showing intervals D and E. *Skolithos* is locally abundant in thick bed just above top of hammer shaft. Head of hammer rests on a conglomeratic bed. Length of hammer shaft 22 cm. In lower part (interval D), bedding consists of alternating thin sandstone–siltstone beds with clayey partings. □B. Detail from lower part of section in A, with sand-filled burrows of *Teichichnus ovillus– Trichophycus* type. □C. Beds from interval E at Hällekis. Just below middle of figure is a distinct horizon, immediately above which is coarse-grained sandstone. Bed below with *Skolithos linearis.*

zia sandstone may have consisted of eroded material reworked during the transgression.

In early works, the sediments in the Mickwitzia sandstone were considered to be of tidal origin (Hadding 1929; Hessland 1955). Some features in the Mickwitzia sandstone have been used to suggest periodic exposure: mass occurrence of stranded medusae (Nathorst 1881a), desiccation mud cracks (e.g., Hadding 1929) and rain-drop imprints (Hadding 1929). Of these, only certain shrinkage cracks are possible, though not probable, indicators of emergence.

There appear to be no sedimentary structures reflecting deposition by tidal mechanisms, such as mud drapes and neap-flood asymmetrical cross-bedding.

Eolian and fluvial processes have been considered important in spreading sediments in such vast sheet-like

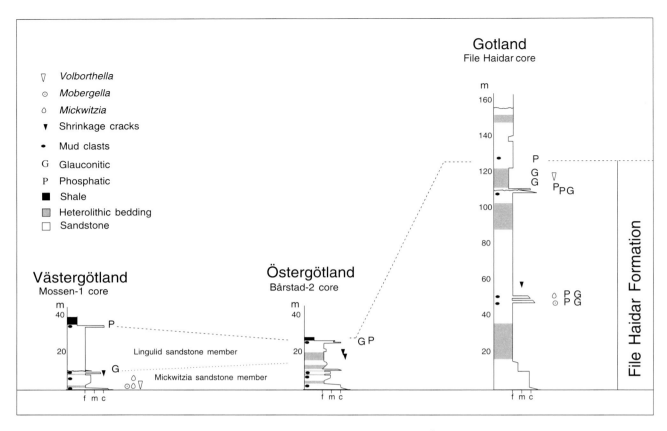

Fig. 18. Simplified graphic logs of sections through the File Haidar Formation in Västergötland, Östergötland and Gotland. Information on lithology and fossils from the sections in Östergötland and Gotland is based on Thorslund & Westergård (1938), Ahlberg (1985) and Eklund (1990).

forms, later modified by the advent of transgression (Dott *et al.* 1986). High proportions of mature quartz-grains in extensive Cambrian cratonic sandstones have led to suggestions of abrasion by eolian processes, which would have been all the more effective in the absence of significant binding by land plants (Dott *et al.* 1986). In the Lower Cambrian Hardeberga Sandstone, Scania, southern Sweden, shallow-water sediments consisting of tidal sediments, including back-barrier sediments, have been recognized (Hamberg 1991). Such sediments appear to be lacking in the Mickwitzia sandstone, which was deposited in shallow water but still some distance away from the shore.

As originally suggested by Martinsson (1965, 1974), storms were probably the main agent depositing sand and silt. The wide development of beds topped by wave-ripples could have been caused by the low gradient of the sea floor, causing vast areas to be affected by storm-wave base (cf. Brenchley et al. 1986). Following a transgressive base, there is a general upward coarsening in terms of proximal and distal parts of storm deposits reflecting shallowing conditions (e.g., Aigner & Reineck 1982; Aigner 1985). The interpretation given here of the Mickwitzia sandstone

envisions a wide epicontinental sea with (low) mud sedimentation during fair-weather conditions. During storms, sediment was transported out from shallow-water sands and deposited as patchy blankets of sand. Much of the mud may have fallen as material put in suspension by storms.

The Mickwitzia sandstone can be compared to other shallow-water, epicontinental, silt- and sandstones in terms of sedimentary structures and trace-fossil content. The Middle Cambrian Paradoxissimus Siltstone of Öland was described by Martinsson (1965). This unit consists of thin siltstone beds separated by muddy layers. Among sedimentary structures in common with the Mickwitzia sandstone are various tool marks, gutter casts (priels), Kinneyian marks and mud clasts. The similarity of trace fossils from the Mickwitzia sandstone to those of the Paradoxissimus Siltstone was noted already by Hadding (1929, p. 150). Trace fossils in common are (names and figure numbers in parenthesis refer to Martinsson 1965): *Palaeophycus sulcatus* (Halopoans); *Teichichnus* and *Trichophycus* isp. (Teichichnians); *Rusophycus eutendorfensis* (Fig. 35E), *Treptichnus* isp. (Fig. 35C); *Dimorphichnus/Monomorphichnus*; and large forms of *Rusophycus*. Martinsson (1965) envisioned a setting below normal wave base with a dominantly clayey deposition inhabited by a coenosis of mud burrowers. Minor influx of silt did not severely disturb these burrowers, though more major influx led to local extinction of the mud-living fauna, including paradoxidid trilobites. The newly deposited silt was colonized by 'halopoans', digging along silt surfaces; this was followed by the return of the mud-dwelling fauna.

Häntzschel & Reineck (1968) reported horizontal to flaser and lenticular bedding of sandstones and siltstones interbedded with shales from the Liassic of Helmstedt, Germany. Sedimentary structures include interference ripples, gutter casts, tool marks, wrinkle marks (? = Kinneyian marks) and Aristophycus-like structures. Trace fossils include *Teichichnus*, *Gyrochorte*, *Koupichnium*, *Curvolithus* and *Rhizocorallium*. Trace-fossil preservation is similar to that in the Mickwitzia sandstone. Forms described as 'Vergabelte Stopfgänge' (Häntzschel & Reineck 1968, p. 29, Pl. 13:1) may be comparable to *Olenichnus* in their mode of preservation. Häntzschel & Reineck (1968) suggested deposition in depths about 40 m, and probably on the low side of this, but also found no supporting evidence for tidal flat setting. Sandy and shaly sediments within the German Bay were given as a modern analog.

Sediment and trace fossils of the Abu Ballas Formation, Egypt, were described by Böttcher (1982). The formation is dominated by claystones, siltstones and fine sandstones with sedimentary structures, including Kinneyian marks. The trace fossils show good preservation and include *Gyrochorte*, *Scolicia* and *Rhizocorallium*. The setting was interpreted as a shallow epicontinental sea with depth probably less than 20 m and with varying salinity conditions (Böttcher 1982).

The Lingulid sandstone and the transition between the Mickwitzia and Lingulid sandstones in Västergötland

The Lingulid sandstone consists largely of fine-grained quartz sandstone, more poorly cemented and thicker-bedded than the Mickwitzia sandstone (Westergård 1931b, 1943; Hadding 1929). Pyrite is common and in the basal beds may form centimetre-sized lumps. Bedding is in the range of a few up to 200 cm, with some of the thickest beds near the top. Thin seams of clayey material separate the beds. Few sedimentary structures are seen. Overall the degree of bioturbation is higher than in the Mickwitzia sandstone.

Trace fossils include poorly preserved *Rusophycus* observed in the basal layers at Hjälmsäter as well as in beds near the top of the Lingulid sandstone at Blomberg on the southern part of Kinnekulle. Also found are *Diplocraterion*, vertical spreite burrows (Jensen & Bergström 1995), and horizontal burrows including laterally shifted spreite. Another trace fossil consists of a large, 35 mm wide, ribbon-shaped trace with arcuate transverse bands, belonging to the *Psammichnites–Plagiogmus* group. Kowalski (1978) included this fossil in the new ichnogenus and ichnospecies *Arcuatichnus wimani*.

Body fossils include the arthropod *Paleomerus hamiltoni* and fragmentary finds of the trilobites *Holmiella* sp. and *Holmia grandis*? (Ahlberg *et al.* 1986; Möller 1987). Recently a mold of a more complete, as yet undescribed, trilobite was found by Mr Holger Buentke (personal communication, 1992). Phosphatic brachiopods also have been found, including *Glyptias favosa* (Westergård 1931b). The Lingulid sandstone has been interpreted as formed in deeper water than the Mickwitzia sandstone, the lower content of clay taken to indicate more intense wave action (Hadding 1927; Westergård 1931b, 1943). Martinsson (1974, p. 245), however, suggested that the differences could depend on formation at a later stage in the sandstone basin.

In older literature, the transition between the Mickwitzia and Lingulid sandstones at Kinnekulle was said to be gradual, and the onset of thick-bedded sandstones (here interval E) was used to mark the base of the Lingulid sandstone (Holm 1901). At Lugnås and Billingen, Linnarsson (1871) reported a sparse conglomerate at the top of the Mickwitzia sandstone. Westergård (1931b) did not find a true conglomerate but reported coarse-grained sediments at the same level. Coarse sandstones, some developed as conglomerates, occur in beds at about 9 m in sections at Hjälmsäter, Hällekis and Mossen. The horizontal continuity of these conglomeratic beds cannot be ascertained, but they occur at much the same level in the sections from Kinnekulle, Lugnås and Billingen. It has been suggested that the Mickwitzia and Lingulid sandstones are separated by a hiatus (Bergström & Gee 1985). The coarse-grained beds may represent several (short) hiati, though Eklund (1990) found no major gaps in acritarch assemblages in the Mickwitzia sandstone in Östergötland, compared to sedimentologically more complete sections on the East European platform. These conglomeratic layers are not deemed useful in marking the top of the Mickwitzia sandstone, as they are weakly developed and may occur at slightly different levels.

In sections at Hjälmsäter and Hällekis, conglomeratic levels are followed by a few decimetres of thin-bedded silt- and sandstones with abundant shrinkage cracks, overlain by a few thick beds with abundant *Skolithos*. In the Mossen-1 core, beds with *Skolithos* are cut by a sharp undulating surface, followed by a coarse-grained layer which, within a few decimetres, turns into the fine-grained more monotonous sediment typical of the Lingulid sandstone. This uneven surface may represent a

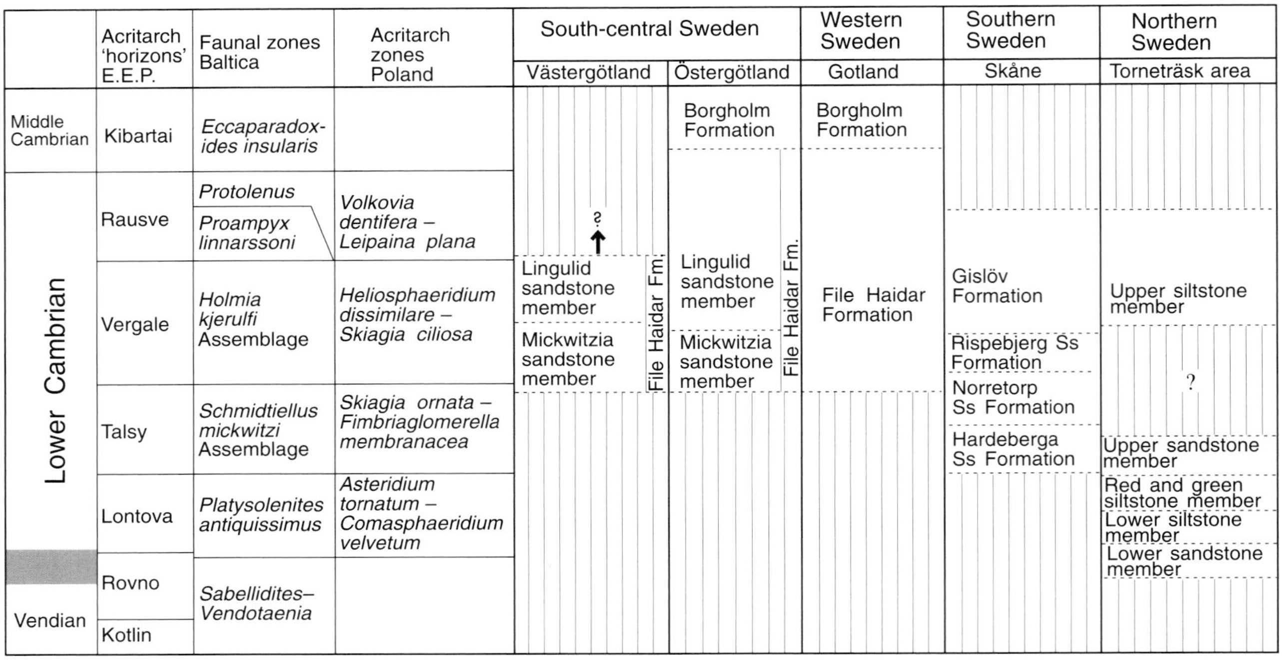

	Acritarch 'horizons' E.E.P.	Faunal zones Baltica	Acritarch zones Poland	South-central Sweden		Western Sweden	Southern Sweden	Northern Sweden
				Västergötland	Östergötland	Gotland	Skåne	Torneträsk area
Middle Cambrian	Kibartai	*Eccaparadoxides insularis*			Borgholm Formation	Borgholm Formation		
Lower Cambrian	Rausve	*Protolenus* / *Proampyx linnarssoni*	*Volkovia dentifera – Leipaina plana*	? ↑ Lingulid sandstone member	Lingulid sandstone member	File Haidar Formation	Gislöv Formation	Upper siltstone member
	Vergale	*Holmia kjerulfi* Assemblage	*Heliosphaeridium dissimilare – Skiagia ciliosa*	Mickwitzia sandstone member (File Haidar Fm.)	Mickwitzia sandstone member (File Haidar Fm.)		Rispebjerg Ss Formation / Norretorp Ss Formation	?
	Talsy	*Schmidtiellus mickwitzi* Assemblage	*Skiagia ornata – Fimbriaglomerella membranacea*				Hardeberga Ss Formation	Upper sandstone member
	Lontova	*Platysolenites antiquissimus*	*Asteridium tornatum – Comasphaeridium velvetum*					Red and green siltstone member / Lower siltstone member / Lower sandstone member
Vendian	Rovno	*Sabellidites–Vendotaenia*						
	Kotlin							

Fig. 19. Tentative correlation of the File Haidar Formation with sections in Scania and the Torneträsk area, northern Sweden, largely based on Moczydłowska (1991). Correlation of Torneträsk Formation based on Bergström (1981) and Jensen & Grant (1992 and unpublished).

more marked zone of non-deposition. A similar sharp border occurs in the File Haidar core on Gotland at a depth of about 390 m and has been interpreted as the result of repeated submergence and emergence (Thorslund & Westergård 1938, p. 17, Fig. 2). Beds with abundant *Skolithos* have typically been assigned as the basal beds of the Lingulid sandstone (Holm 1901; Westergård 1931b). They are here reinterpreted to belong to the Mickwitzia sandstone, based on their position below a surface that may mark a more substantial break in sedimentation. The boundary, or rather transitional interval, as applied here is set within the upper part of interval E, in glauconitic cross-bedded layers that upwards give way to more consistently fine-grained beds (Figs. 3, 15A, C). A similar approach was favoured by Möller (1987).

The File Haidar Formation and the stratigraphical position of the Mickwitzia sandstone member

The Mickwitzia sandstone is defined on lithological characteristics. The name-giving brachiopod is reasonably common in Västergötland, mainly as fragments, but only in a few beds in the lower part. Though not meeting the current recommendation (Hedberg 1976) that fossil taxa should not be used as names for lithostratigraphic units,

there are reasons of continuity for adhering to the old name, treating it without a rigid biological connotation (Bengtson 1980). The Mickwitzia sandstone was designated a member of the File Haidar Formation by Bergström & Gee (1985), but it has not been formally designated. The type section of this formation is a boring at File Haidar, Gotland, described by Thorslund & Westergård (1938). The formation is dominantly siliciclastic and contains few body fossils. The File Haidar Formation was formed as part of the extensive early Cambrian transgression, considered to have been coupled with the early development of the Iapetus Ocean (e.g., Piper 1985; Hagenfeldt 1994). In the area under discussion, the transgression is supposed to have progressed in a southeast–northwest direction with reference to today's orientation (Thorslund 1960; Martinsson 1974; Hagenfeldt 1989). Within the Mickwitzia sandstone in Västergötland, two more pronounced cycles of shallowing can be seen, one stretching through intervals A, B and C; the other, beginning with an intraformational conglomerate at the top of C, spanning interval D and part of interval E. There is a general similarity in lithological succession for the sediments in Västergötland, Närke and Östergötland. The vertical extent of the Mickwitzia sandstone increases to the east (Fig. 18). On the other hand, the thickness of the Lingulid sandstone decreases toward the east; in Västergötland the thickness of the Lingulid sandstone Member is about 25 m, in Närke about 7–10 m and in

Östergötland less than 5 m. In eastern Sweden, the File Haidar Formation is known only from cores. These include cores from Öland (Borghamn, Westergård 1929; Böda Hamn, Waern 1952; Hessland 1955), Gotland (File Haidar, Hedström 1923; Thorslund & Westergård 1938; Ahlberg 1989; När, Ahlberg 1989), Gotska Sandön (Thorslund 1958), and the South Bothnian Sea (Västra banken and Finngrundet, Thorslund & Axberg 1979).

A substantial part of the Lower Cambrian in Scandinavia is covered by four trilobite-based biozones: the *Schmidtiellus mickwitzi*, *Holmia inusitata*, *Holmia kjerulfi* and *Proampyx linnarssoni* Biozones, based on the succession in the Mjøsa area, Norway (Bergström 1981; Ahlberg 1984; Ahlberg *et al.* 1986; Bergström & Gee 1985). Recently the *H. inusitata* Biozone has been incorporated in the *H. kjerulfi* Assemblage Zone (Moczydłowska 1991, Ahlberg & Bergström 1993). Using macrofossils, the Mickwitzia sandstone has been correlated to the *Schmidtiellus mickwitzi* Biozone (Bergström 1981; Ahlberg 1984; Bergström & Gee 1985). This zone is typified by the trilobites *Holmia mobergi*, *Holmia* cf. *mobergi*, *Schmidtiellus mickwitzi* and *Kjerulfia? lundgreni* (Ahlberg 1984). In the absence of trilobites, correlation of the Mickwitzia sandstones has largely been based on the brachiopod *Mickwitzia monilifera* and the problematicum *Mobergella* sp., found together with trilobites indicative of the *Schmidtiellus mickwitzi* Biozone in Estonia and Norway, respectively (Bergström 1981).

However, both *Mobergella* and *Mickwitzia* may have a stratigraphic range exceeding the *Schmidtiellus mickwitzi* Biozone (Hagenfeldt 1989; Moczydłowska 1991). These fossils appear to be facies controlled and are restricted to levels with intraformational conglomerates. Brangulis (1985) listed *Mobergella* from the Vergale and Rausve horizons and *Mickwitzia monilifera* from the Rausve horizon in Latvia. The problematicum *Spatangopsis costata* is known from the Mickwitzia sandstone in Västergötland and Närke and through one specimen from the Lükati Sandstone in Estonia. Its correlative value is doubtful, since it may be an inorganic structure. Fragments of a trilobite identified as *Holmiella* sp. were reported by Ahlberg *et al.* (1986) from a section at Hällekis. This find has been assigned to the Mickwitzia sandstone (Bergström 1981; Ahlberg 1984) or the basal part of the Lingulid sandstone (Ahlberg *et al.* 1986). Comparison with the lithology log in Ahlberg *et al.* (1986) indicates that this occurrence is in the Mickwitzia sandstone member. Fragments of *Holmia grandis?* in the Lingulid sandstone have been taken to indicate correlation with the *Holmia kjerulfi* Assemblage Zone (Ahlberg *et al.* 1986). The body-fossil control of the Mickwitzia sandstone suggests that it is not younger than the *Holmia kjerulfi* Assemblage Zone. The presence of *Mobergella* in the basal part could indicate correlation with the *Schmidtiellus mickwitzi* Biozone, though this is hampered by poor preservation of the specimens and uncertainties regarding the species identity of these *Mobergella*.

Acritarch work on the Mickwitzia sandstone in Västergötland is still in its infancy. From basal layers of the Mickwitzia sandstone, Vidal (1981b) and Moczydłowska & Vidal (1986) reported 18 taxa, noting that several of these also were found in the Norretorp Sandstone; correlation was also made to unit 1α of Mjösa. Hagenfeldt (1989) considered the presence of *Skiagia ciliosa* and *Micrhystridium notatum* in the Mickwitzia sandstone to indicate the *Holmia kjerulfi* Assemblage Zone, or younger, and Moczydłowska (1989, 1991) found the acritarchs to indicate the Vergale horizon, corresponding to the *Holmia kjerulfi* Assenblage Zone. Eklund (1990), working on core material from Östergötland, found a sequence of acritarchs comparable to assemblages that in the East European Platform indicate the Vergale horizon. Acritarchs from the Lingulid sandstone in Östergötland indicated correlation with the Rausve horizon, and the top part was even suggested to be Middle Cambrian. Though the sandstones form a condensed sequence, comparison with more extensive sections on the East European Platform showed no major hiati (Eklund 1990).

The main part of the Mickwitzia sandstone member in Västergötland may thus belong to the *Holmia kjerulfi* Assemblage Zone, while the basal layers could possibly belong to the *Schmidtiellus mickwitzi* Biozone (Fig. 19).

Ichnotaxonomy, ichnostratigraphy and metazoan evolution

The use of a binomial naming system for trace fossils is firmly established among ichnologists. Only two levels of ichnotaxa are in common use; ichnogenus and ichnospecies, following Bromley (1990) abbreviated igen. and isp. Today, most students accept ichnogenera as based on morphologic traits reflecting behaviour at a high level of significance (e.g., Fürsich 1974a; Pemberton & Frey 1982). Bromley (1990) listed four major criteria that are widely used in defining ichnotaxa: general form, branching, burrow fill and structure, and burrow boundaries. Generally inappropriate features to base ichnogenera upon include: size and crowding, preservational differences, environment in which the trace is made, stratigraphic level and regional occurrence (see discussions in Bromley 1990; Magwood 1992; Pickerill 1994). In most cases, it is also inappropriate to base ichnogenera on a specific producer; many ichnogenera are known to be made by animals of several phyla. Still, because of the highly varied nature of trace fossils, no standard tool exists upon which to define ichnogenera; some, such as *Zoophycos* are widely defined whereas others are more narrowly defined. As a rule, the more details a trace fossil reveals of its originator, the greater the possibility and temptation to erect narrowly defined ichnogenera. An exemple are arthropod track-ways where ichnogenera are often based on such features as number and pattern of claw impressions (e.g., Walter 1984). Not surprisingly, the literature is flooded with names, most of which have never been used outside their type area. However, in general a workable scheme of ichnogenera has emerged, the foundation of which was laid by Häntzschel (1975), and with refinements by Fürsich (1974a, b), Pemberton & Frey (1982) and D'Alessandro & Bromley (1987) and others. Following Fürsich (1974a, p. 956) ichnospecies are defined upon 'morphological features resulting from behavior of a low degree of significance'. In *Diplocraterion*, Fürsich (1974a) recognized a vertical U-tube with a spreite as an ichnogeneric characteristic, and burrow outline and direction of growth as bases for ichnospecies.

In some ichnogenera features relating directly to details in morphology of the producer have been used in diagnosis. The rational for this has often been the possibility to assign the trace to a given producer and thereby make it of potential stratigraphic use. Especially notable is the stratigraphic use of species of *Cruziana,* which has proved helpful in Lower Palaeozoic Gondwana stratigraphy (e.g., Crimes 1968; Seilacher 1970, 1992b). Even so, trace fossils will always give rather restricted morphologic information, and the risk of convergent morphologic details is high. Magwood & Pemberton (1990) report the occurrence of species of *Cruziana* from the Lower Cambrian of Alberta, Canada, that in Europe are used as index fossils for the Lower Ordovician (but see Seilacher 1994b).

The most objective approach to ichnotaxonomy is to base ichnotaxa solely on the preserved morphology. However, the degree to which taphonomic factors influence morphology is problematic and have been little considered. Preservation of discrete trace fossils is largely dependent on properties of the sediment and the boundaries between different lithologies (Fig. 20). Thus, preservation of such features as surface sculpture is dependent on properties of the sediment. Morphological details may form after the trace fossil was produced, such as transverse ribs on *Gyrolithes polonicus* in the Mickwitzia sandstone. Morphological structures may also be degraded and lost during diagenesis. MacNaughton & Pickerill (1995) termed this taphoseries and gave as an example loss of wall-lining and back-fill, leading to the morphology of different ichnotaxa. Though it may be argued that interpreting the diagenetic history introduces an element of subjectivity, a basic sorting out of taphonomic factors is essential.

Trace fossils may pass directly from one morphology into another (compound specimens of Pickerill 1994). This is probably more common than would appear from a reading of the literature. Pickerill & Narbonne (1995) advocated naming each form of a compound specimen

separately with the dominant form expressed first. Compound specimens are direct evidence that two different trace fossil morphologies were made by the same producer, though this is complicated by simultaneous and sequential use by different animals. However, if a direct continuation forms only a small part of the total area covered by two trace fossils, this connection could easily be lost or overlooked. Occasionally, it is also possible to demonstrate that different trace fossils had the same producer by identifying tell-tale features of the animal. This includes distinctive claw patterns in arthropods. Even where such features exist, it is possible that otherwise widely different animals may leave similar signatures.

There are taphonomic reasons to expect that assemblages of well-preserved trace fossils will represent the activity of a minor part of the burrowing community. The preservational potential of surface traces is small, even under conditions where the obliterating effect of bioturbation is low. The trace-fossil record in marine settings will therefore be highly biased towards traces made well below the sediment–water interface and typically at a late stage in trace fossil successions (e.g., Bromley 1990). These traces will reflect the activity of infaunal burrowers or large surface-ploughing animals. Hertweck (1972) studied traces from shallow nearshore marine sediments at Sapelo Island, Georgia, USA. Of 268 reported animals species, 40 produced distinct traces; of these about half had the potential to be preserved, with six species of trace makers dominating (Hertweck 1972). There is thus reason to suspect that the trace fossils in the Mickwitzia sandstone will reflect the activity of only a minority of the animals present. An example are centimetre-sized, U- to L-shaped burrows with a passive fill, including some with spreite. These include *Trichophycus, Teichichnus* and probably some examples of *Palaeophycus imbricatus*. It is also most likely that most of the funnel-shaped trace fossils can be tied to horizontal burrows. These include *Rosselia socialis*, which appears to be the vertical part of *Palaeophycus* and perhaps of *Zoophycos* isp. The same is probably true for *Monocraterion*, including the type species, *M. tentaculatum*. Especially revealing are *Monocraterion* preserved in the heterolithic level, where they occur on the bases of sandstones together with abundant *Palaeophycus*. Because of compaction, direct continuations are rarely seen, but they can be inferred from repeated outlets from the vertical funnels and similarities in size. Direct continuations are also seen between *Palaeophycus imbricatus*, *Scotolithus mirabilis* and *Phycodes* cf. *curvipalmatum*. Another example of a tunnel system is *Olenichnus*, which differs in the type of preservation and in forming more winding paths (Fig. 21). All of these dwelling or dwelling–feeding structures could have been made by the same type of animal

Thus, application of trace fossils in biological and evolutionary studies must be strongly tempered.

For example, how should the increase in ichnodiversity at the Precambrian–Cambrian boundary be interpreted? Is the increase in morphological diversity linked to the activity of a few taxa or does it reflect greater variety of animal body plans and behaviour?

Attempts to classify trace fossils into functional groups is problematic. Often a trace fossil reveals only part of the full original morphology, and there are typically multiple and commonly poorly known functions of modern traces. Trace fossils give us some of the earliest examples of probable metazoan activity. However, due to the taphonomy of trace fossils, this record will be strongly, or exclusively, biased towards the activity of burrowing animals. If, as commonly assumed, the earliest stages of bilaterian metazoans were surface-creeping rather than burrowing forms (e.g., Clark 1964, Bergström 1990) it is not surprising that there is no record of their activity. Similarly, if Vendian fossils such as *Parvancorina* and *Spriggina* are interpreted as limbed arthropods, the lack of surface locomotion trackways in sediments yielding these forms should not be used to infer a swimming mode of life. Surface traces are extremely rarely preserved in marine sediments of any age, let alone from animals of such relatively small size. For example, it must be questioned if there are any convincing examples of trilobite (or other arthropod) trace fossils produced in locomotion on the sea floor. Forms such as *Diplichnites* are in most cases clearly undertrack fallout of *Cruziana*, which is not a repichnion.

The Precambrian–Cambrian boundary

Trace fossils have emerged as important tools in correlation at the Precambrian–Cambrian boundary, not least because they are common in many clastic, otherwise fossil-poor successions (e.g., Seilacher 1956; Crimes 1975a, 1987, 1989, 1992, 1994; Alpert 1977; Fedonkin 1985; Walter *et al.* 1989). In numerous sections world-wide a similar order of first appearance of ichnogenera has been reported, with generally simple horizontal traces in the Vendian and an increase in complexity and diversity in the Cambrian (e.g., Banks 1970; Paczesna 1986; Narbonne *et al.* 1987; Walter *et al.* 1989). Crimes (1987) recognized three trace-fossil zones at the Precambrian–Cambrian boundary based on the incoming of new ichnotaxa. Zone 1, of late Vendian age, includes mainly horizontal forms such as *Gordia*, *Harlaniella* and *Palaeopascichnus*. Zone 2, of early Tommotian age, includes more complex forms such as *Phycodes*, while Zone 3, of late Tommotian to early Atdabanian age, is marked by a major increase in diversity, including traces such as *Cruziana* and *Rusophycus* typically attributed to trilobites. Reports of trace fossils older than the Vendian occur, but upon critical analysis they have turned out to be inorganic, misdated or otherwise highly doubtful (Cloud 1968; Byers 1982; Bergström 1990; Fedonkin & Runnegar 1992; Runnegar 1992; Hofmann

1992; Crimes 1994). Crimes (1992) recently listed numbers of ichnogenera across the Precambrian–Cambrian transition. The number was given as about 30 in the Redkino and Kotlin, with a rise to about 40 in the Rovno and about 60 in the Tommotian. The vagaries of ichnotaxonomy were acknowledged, and though only the most obvious synonyms had been taken out of this list, Crimes (1992) felt that the general picture of increased diversity remains.

A more serious problem is the difficulty in interregional calibration of the appearance of trace fossils, due to paucity of, and preservational/diagenetic control of other stratigraphic indicators at the Vendian–Cambrian. This is especially true for comparisons between terrigenous regions and the largely carbonatic Siberian platform. Thus, equivalents to the Siberian Tommotian and Atdabanian stages, which have come be to used globally, are in other regions open to dispute (e.g., Moczydłowska & Vidal 1988).

Some ichnotaxa have been considered to be restricted to certain zones. There has been little discussion on how the restricted ichnogenera are unique and what they signify. Taxa considered restricted to the Redkino include *Buchholzbrunnichnus* from Namibia, *Bunyerichnus* from Australia and *Intrites*, *Medvezhichnus*, *Nenoxites*, *Vendichnus*, *Vimenites* and *Yelovichnus* from the White Sea Area of Russia. Of these, *Bunyerichnus* is now considered to be inorganic (Jenkins *et al.* 1981), or to be a body fossil of Ediacaran type (Glaessner 1984).

The trace-fossil interpretation of *Buchholzbrunnichnus* is also questionable. *Intrites punctatus* Fedonkin, 1980, consists of rows or irregular groups of small knobs in hyporelief, the formation of which is uncertain. Fedonkin (1980) suggested that it was formed by an animal repeatedly probing across a sediment boundary. It has also been compared with *Bergaueria* (Crimes 1987) or *Protolyellia* (Bergström 1990, p. 5). *Vendichnus vendichnus* Fedonkin, 1981, is a rare, irregularly bilobate form possessing short striae. It was compared with resting traces made by arthropods, especially *Ixalichnus* Callison, 1970, and tentatively suggested to be made by a bilaterally symmetrical animal (Fedonkin 1981). *Vimenites bacillaris* Fedonkin, 1980, and *Nenoxites curvus* Fedonkin, 1976, are meandering forms, of which the latter has been suggested to show signs of peristaltic movement (Fedonkin 1985, p. 123). *Yelovichnus gracilis* Fedonkin, 1985, has been described as a tightly guided meandering trace similar to Phanerozoic forms generally found in flysch deposits. However, no illustrated specimen clearly demonstrates a continuous meander. *Medvezhichnus pudicum* Fedonkin, 1985, has a circular cross-section and a narrow basal sinusoidal ridge (Fedonkin 1985).

Of traces supposedly restricted to the Kotlin stage, *Harlaniella podolica* is of special interest. *Harlaniella* has been reported from the Kanilov Suite of Ukraine (e.g., Palij

1976), Member 1 of the Chapel Island Formation on Newfoundland (e.g., Narbonne & Myrow 1988), and the Lublin Formation of southeastern Poland (Paczesna 1986). This rope-shaped trace has been interpreted as a feeding or grazing structure of a worm-shaped organism (Fedonkin 1985). *Harlaniella* is usually considered to be restricted to the latest part pf the Vendian, and its range has defined a *Harlaniella podolica* Zone in the Precambrian–Cambrian stratotype region on Newfoundland (e.g., Narbonne & Myrow 1988). Signor (1994) extended the range of *Harlaniella* to the Cambrian by reporting a second species, *Harlaniella confusa*, from the Lower Cambrian of the Wood Canyon Formation, California. This form differs from *H. podolica* by '... the density of the spiral, the occasional reversals in the direction of the spiral, and the apparently random orientation of the spiral within the rock.' (Signor 1994, p. 320). The three-dimensional morphology of *Harlaniella* has not been demonstrated. At least superficially similar trace fossils occur in Phanerozoic flysch deposits, including *Helicorhaphe tortilis* Książkiewicz, 1970, and *Helicoichnus cylindricus* Yang Shipu, 1986. *Helicorhaphe* and *Helicoichnus* are known from the Tertiary and Cretaceous respectively (Książkiewicz 1977; Yang *et al.* 1986). Furthermore, until the corkscrew shape of *Harlaniella* has been documented in three dimensions, its similarity to certain forms of *Phycodes pedum* is noticeable (compare Pl. 7b of Crimes *et al.* 1977 with Pl. 1:1 of Kirjanov 1968)

From the above it can be seen that Vendian trace fossils are comparable to Palaeozoic traces or have a poorly known morphology. Traces typical of deep-water environments through most of the Phanerozoic have in Cambrian deposits been reported from shallow-water settings (Crimes & Anderson 1985; Hofmann & Patel 1989; Crimes & Fedonkin 1994). The interpretation of purported Vendian meanders such as *Palaeopascichnus* and *Yelovichnus* is still problematic, since there are no documentation of continuous meanders. The reasons for making patterns such as meanders and spirals include: more effective use of patchy food availability, high individual density, and reducing the risk of detection by predators (e.g., Kitchell 1979). Crimes (1992) suggests that high levels of competition forced some forms away from their favored sediments to occupy areally restricted, quiet, muddy substrates where intricate patterns were developed. These animals were then driven out to deeper water when competition further increased (Crimes 1992). The timing of colonization of deep-water environments is uncertain owing to poor availability of corresponding strata, but diverse deep-water assemblages of traces are known from, e.g., the Ordovician–Silurian of eastern and western Canada (Pickerill 1980; Pickerill *et al.* 1987, 1988), and the Silurian of Wales (Crimes & Crossley 1991).

The producers of the Vendian trace fossils remain unknown. Most, and probably all, Vendian trace fossils, are shallow endogenic burrows. This can be seen by the common preservation in negative hyporelief, as has been pointed out by Seilacher (1992a). This in itself gives little information; burrowing is often, following Clark (1964), stated to be restricted to segmented coelomate animals, but a coelome is not a prerequisite for a burrowing life (e.g., Willmer 1990). An important point is whether the advent of more complex three-dimensional traces represents the rise of organisms with new morphologic abilities or was dependent on new requirements caused by increased competition. Simple vertical traces have been reported from the Redkino of the White Sea area, viz. *Skolithos declinatus* Fedonkin, 1985. Reports of the rosetted trace fossils *Asterichnus* and *Stelloglyphus* in the Vendian Mogilev–Podolskaja Series of Podolia, Ukraine (Gureev 1984; Gureev *et al.* 1985) is of interest, but needs closer documentation.

Recently a stratotype for the Precambrian–Cambrian boundary at Fortune Head, Burin Peninsula, southeastern Newfoundland (Narbonne *et al.* 1987), has been selected. The base of the Cambrian is defined to coincide with the lowest occurrence of *Phycodes pedum* in this section (Landing *et al.* 1988). The base of the Cambrian is 2.4 m above the base of the Chapel Island Formation at Fortune Head, 0.2 m above the highest occurrence of *Harlaniella*, a trace considered restricted to, and characteristic of, the upper Vendian (but see above).

The underlying rational for using trace fossils in stratigraphy at the Precambrian–Cambrian boundary must be that the developments seen in trace patterns reflect the appearance of metazoans capable or inclined to produce a new behaviour pattern, or faced with conditions under which such patterns were advantageous. Trace fossils are notorious for convergence in patterns produced by unrelated animals of great morphologic difference. Furthermore, behaviour is not tightly locked into discrete patterns but will show transitions, as exemplified by *Rusophycus* and *Cruziana*. A first appearance of readymade *P. pedum* is not to be expected, but rather a transition from less to more typical *P. pedum*. *P. pedum* may be seen as an accumulation of building units which in themselves could be considered short, curved, *Palaeophycus* or *Planolites*. It is of prime importance that the intermediate phases between a horizontal burrow and *Phycodes pedum* be studied and that it is decided where to put the 'lower' morphological limit of *Phycodes pedum*. As an example of the morphological range, specimens such as those illustrated by Nowlan *et al.* (1985, Fig. 5), are little different from *Palaeophycus*. At the other extreme, this type of burrow grades into the long chains of segments in *Treptichnus* and '*Hormosiroidea*' (Fig. 23), or in the case of short closely set branches, to *Harlaniella*. As discussed in the sections on *Phycodes* and *Treptichnus*, *Phycodes pedum* may more properly be assigned to *Treptichnus*.

The use of a single ichnogenus to infer stratigraphic level at the Precambrian–Cambrian boundary is doubtful, given that most ichnotaxa are poorly understood, and that trace fossils such as *Treptichnus pedum* are dependant on mode of preservation, and thus sedimentary environment, to be easily recognizable. Of greater utility is the use of assemblages of trace fossils. One such assemblage that appears low in the succession in several areas consists of *Treptichnus*, *Gyrolithes* and *Curvolithus*. This type of assemblage was found by Jensen & Grant (1992, and work in progress) from low in the middle sandstone of the Tornetträsk Formation, northern Sweden, which on the basis of the problematic medusoid *Kullingia* had previously been thought to be Vendian.

The Mickwitzia sandstone

The Mickwitzia sandstone contains trace fossils first encountered in the third trace-fossil zone of Crimes (1987). These include *Diplocraterion*, *Rusophycus* and *Cruziana*. In sections worldwide, *Rusophycus* and *Cruziana*, appear slightly below the first trilobite body fossils; *Cruziana* appears in beds that have been correlated near the Tommotian–Atdabanian boundary (Crimes 1987), while *Rusophycus* in Poland may appear in the Lontova horizon (Fedonkin 1977), correlative with, or older than the Tommotian Stage (Moczydłowska & Vidal 1988). Crimes (1975a, 1987) noted that many early *Rusophycus* have transverse scratches, a feature thought to result from ineffective musculature. Seilacher (1960) considered the predominance of rusophyciforms to cruzianiforms in '*Cruziana dispar*' as a primitive character compared to Ordovician forms where band-like shapes are more common. For the Middle Cambrian *Olenoides serratus*, Whittington (1980) found it unlikely that there was enough forward pull to achieve a simultaneous digging activity and forward progression. This may have been generally true for Cambrian trilobites. The arthropod trace fossil *Rusophycus dispar* has been reported together with trilobites indicative of the *Schmidtiellus mickwitzi* Biozone in the Mjøsa district and Scania (Bergström 1981). The identification of *Rusophycus dispar* as made by an olenelline trilobite is reasonably firm (Bergström 1973a; Jensen 1990, this work). Its correlative value (Bergström 1981) is, however, more doubtful. Many, or most of the olenellid trilobites were probably capable of producing *Rusophycus* burrows. Preservation of their burrows would be highly dependant on suitable sediments, which may govern the occurrence of *Rusophycus dispar* in the Mickwitzia sandstone and in the Norretorp Sandstone, Lower Cambrian of Scania, southern Sweden. A future prospect would be to search for differences in claw pattern, size and outline

among supposed olenelline burrows in the Baltic area. *Rusophycus parallelum* Bergström, 1970, was reported from the Hardeberga Sandstone in Scania (Bergström 1970). It consists of shallow burrows with parallel sides and imprints of two distal claws (Bergström 1970). With only two specimens known, differing from each other in the V-angle, it is difficult to compare this form with *Rusophycus dispar* and *Cruziana rusoformis*, the variation within which is considerable.

Among trace fossils in the Mickwitzia sandstone with possible stratigraphic value, *Syringomorpha* is especially noticeable. Alpert (1977) listed *Syringomorpha* as a trace indicative of early Cambrian age, and Fedonkin (1979) grouped it with traces first appearing in the Atdabanian. This was recently put in doubt by a report of *Syringomorpha nilssoni* from the upper part of the Vendian McManus Formation of North Carolina (Gibson 1989). This trace has subparallel adjoined burrows with an outline broadly similar to that of *Syringomorpha*. However, it differs by having an orientation that is parallel to the bedding. The North Carolina trace fossils were compared to *Oldhamia* (Seilacher 1992a, p. 610) but this was unacceptable to Crimes (1994, p. 115). They could be compared to *Agrichnium* Pfeiffer, 1968. Chiplonkar & Badve (1970) reported *Syringomorpha* sp. indet. from the Cretaceous Bagh Beds of India, but without giving the orientation of the trace the identification is dubious. At present, well-documented occurrences of *Syringomorpha* (Hadding 1929; Lendzion 1972; Orłowski 1989; this study) are known only from the Baltic area and the Holy Cross Mountains where it is restricted to a relatively narrow interval of the Lower Cambrian.

Also, *Halopoa* has been suggested to be a short-ranging form, not extending beyond the Cambrian (Crimes & Crossley 1991, p. 61). However, *Halopoa* is here considered a heterogenous taxon containing forms of *Palaeophycus*, *Phycodes* and *Scotolithus* (see description of *Palaeophycus imbricatus*). The type species, *Halopoa imbricata*, is probably identical to *Fucusopsis* or *Palaeophycus*

Notable is the occurrence of different types of *Zoophycos* in the Mickwitzia sandstone. *Zoophycos* is generally considered to appear with flat nonspiral forms in the Ordovician, e.g., in the Dersish Formation, Iraq (Seilacher 1964), and the Trenton Group, Canada (Fillion & Pickerill 1984), while spiral forms seem to appear only in the Devonian. Reports of *Zoophycos* in Cambrian sediments are very rare and poorly documented. Alpert (1977) listed *Zoophycos* from the Lower Cambrian White–Inyo Mountains, California, but judging from specimens illustrated by Alpert (1974a, 1976a) they consist of arched burrows, arranged in a radiating or progressively shifted manner resulting in an elongate to irregular outline. In the latter case there is resemblance with *Fuersichnus* Bromley & Asgaard 1979, which consists of bow or J-shaped burrows that may form a loosely packed tongue or ear-shaped spreite, and have transverse striation or knobby ornamentation (Bromley & Asgaard 1979). In the case of a radiating arrangement, comparison can also be made with radiating burrows, possibly *Bifasciculus*. Magwood & Pemberton (1988) reported 'possible *Zoophycos*-like fragments' in float of the Lower Cambrian Gog Group of Alberta, Canada. Finally, Bryant & Pickerill (1990) reported cf. *Zoophycos* from the Buen Formation of Greenland, based on a single specimen. The Mickwitzia sandstone specimens may thus be among the oldest reports of *Zoophycos*. This squares with the notion of

Crimes (1992), that by Atdabanian time the main Phanerozoic trace-fossil lineages had already evolved. A migration to deeper water for many types of ichnotaxa with time has been compared to similar trends in benthic communities (e.g., Jablonski *et al.* 1983; Sepkoski 1987). Bottjer *et al.* (1987) recorded the occurrences of *Zoophycos* through the Phanerozoic and found a trend with increasing restriction to deeper-water sediments in the Mesozoic. Specimens in the Mickwitzia sandstone document the occurrence of *Zoophycos* in shallow-water sediments from the Lower Palaeozoic.

Preservation of trace fossils

Most trace fossils in the Mickwitzia sandstone are typical of the *Cruziana* ichnofacies of traces made near the sediment–water interface by animals dwelling and/or feeding (Frey & Seilacher 1980). The preservation of trace fossils in the Mickwitzia sandstone is intimately connected with episodic sedimentation. This is clearly seen in traces exhibiting fluting formed when the trace was exhumed and recast. Open burrows thus became rapidly filled and may possess a lag of coarse sediment. These open burrows and burrow systems often served as traps for sediment otherwise not deposited on the sediment surface. Some animals were capable of reworking the sediment by pressing it to the walls or raising the level of the burrow, now reflected in spreite of *Teichichnus*, *Trichophycus*, and perhaps *Diplocraterion*. The upper part of a burrow system was most easily filled, whereas deeper parts, especially of longer horizontal burrows, largely remained open and subsequently collapsed; this is the case in *Olenichnus* (Fig. 20B). In dense populations of large open burrows, passive sedimentation results in substantial trapping of sediment. Large burrow systems trapping substantial amounts of storm-deposited sediments, especially such made by digging crustaceans, are referred to as tubular tempestites (Wanless *et al.* 1988; Tedesco & Wanless 1991). Repeated filling and recolonization may result in a total alteration of the sediment (Wanless *et al.* 1988). Isolated lenses of sediment in the Mickwitzia sandstone are also common with the trilobite traces *Rusophycus dispar* and *Cruziana rusoformis*. The material filling up these traces often consists of a lens of sediment pressed onto the bottom of a

more extensive bed by compaction. There have been different opinions on the preservation of this type of trace. Some favour its formation as an open furrow in mud, later cast by sediment (Crimes 1975b; Baldwin 1977). Others advocate an intrastratal formation with the trace being preserved along the interface of a mud and sand layer (Seilacher 1970, 1985). Traces formed intraforma-

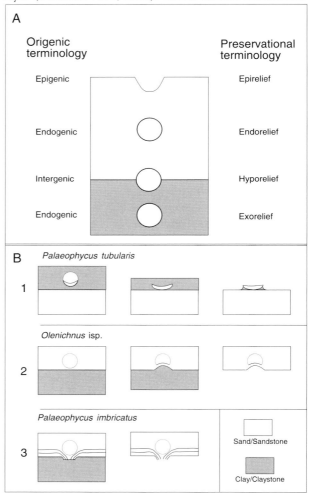

Fig. 20. □A. Trace-fossil terminology. Origenic terminology describes relation to strata at the time of trace production. Preservational terminology describes relation to strata (casting medium) in which trace is preserved. Note that there is no forced coupling between origenic and preservational terminology. □B. Examples of influence of sediment condition on trace fossil morphology and preservation. In *Palaeophycus imbricatus* laminae are deflected by the burrowing animal.

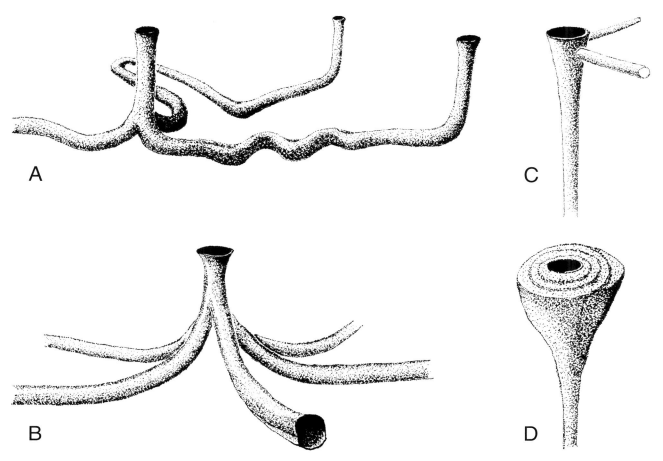

Fig. 21. Interpretative reconstructions of selected trace fossils. □A. *Olenichnus* isp. □B *Scotolithus mirabilis* □C. *Monocraterion tentaculatum.* □D. *Rosselia socialis.*

tionally may later be exhumed and cast, leaving a laminated fill (Goldring 1985). In the Mickwitzia sandstone, an intraformational fill is suggested by (1) specimens with sharp basal structures but scoured sides, (2) rare specimens with mottled sand reflecting dug-up mud, and (3) indirect indications from specimens associated with burrows showing response to sediment influx (Jensen 1990). The intrusion of sediment during storms resulted in a sedimentary cover that was exploited by animals producing *Palaeophycus imbricatus*, *Phycodes* cf. *curvipalmatum* and *Scotolithus mirabilis*.

More vertical to inclined traces are met in interval C, with *Syringomorpha*, *Rhizocorallium* and *Diplocraterion*, in addition to some *Rusophycus* and *Palaeophycus imbricatus*. This may be considered a *Cruziana* Ichnofacies deposited under higher-energy conditions, giving aspects of the *Skolithos* Ichnofacies. A purer *Skolithos* Ichnofacies is found in interval E, which has abundant *Skolithos*. Dense occurrences of *Skolithos* are often found in storm deposits, notably hummocky cross-stratification (Droser 1991), which may also be the case in interval E. Dense occurrences of *Rhizocorallium* immediately above the

basal conglomerate and near the top of level C are probably connected with transgressive conditions with erosion or little or no sedimentation. This may form an example of the *Glossifungites* Ichnofacies with firm but unlithified sediments (cf. Frey & Seilacher 1980)

In addition to sediment consistency, the generally low intensity of bioturbation in level B and D was favourable to trace-fossil preservation. A possible cause of this would be low oxygen levels. Circulation within a wide shallow sea is highly influenced also by small-scale topography, and large areas of stagnant bottom conditions are to be expected (e.g., Hallam 1981). Storms could have been a major agent to cause oxygenated water to enter otherwise restricted environments. Single storms, may however, not be enough to break oxygen-stratified waters (Steimle & Sindermann 1978), making seasonal control more likely. Extensive bulldozing of shelf sediments appeared in the later part of the Palaeozoic (Thayer 1983). Besides trilobites, the major early Palaeozoic sediment disturbers may have been the animal(s) forming traces of *Psammichnites–Plagiogmus–Didymaulichnus*, specimens of which may reach considerable dimensions. Traces belonging to this

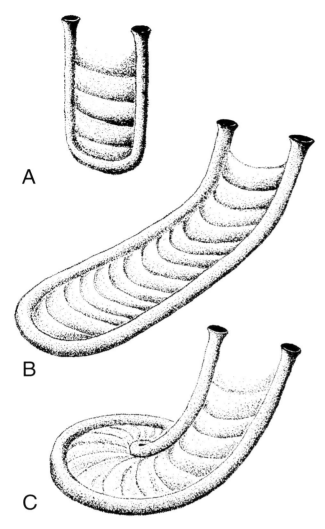

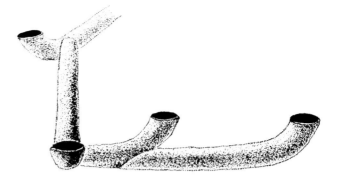

Fig. 23. Interpretative reconstructions of suggested relationsship between the ichnotaxa '*Hormosiroidea*' isp. (right part) and *Treptichnus*.

Fig. 22. Interpretative simplified reconstructions of trace fossils that form a continuous morphological series. □A. *Diplocraterion parallellum.* □B. *Rhizocorallium jenense* □C. *Zoophycos* (*Rhizocorallium?*) isp.

Terminology. – Terms to describe the original position of traces in relation to the sediment follow that of Chamberlain (1971); using epigenic for surface traces (corresponding to exogenic in the terminology of Seilacher), endogenic for traces made within the sediment, with intergenic as a special case for traces made along a sedimentary interface (Fig. 22A). Preservational terms, more or less synonymous to the relation between the trace and casting medium, follow Seilacher (1964), as expanded by Chamberlain (1971, Fig. 3). The scheme developed by Martinsson (1965, 1970) is in wide use, however, I find the definition of Martinsson's terms to be a mixture of origenic (relation to sediment at time of trace formation) and preservational conditions. Though in a purely descriptive sense it parallels Seilacher's scheme, unfortunate origenic implications were added – 'both hypichnia and epichnia are exogenic (impressed or otherwise marked as tracks, trails, etc, upon muddy and sandy surfaces, respectively' (Martinsson 1970, p. 326). It was also assumed that the terms endogenic and exogenic became superfluous. This is misleading in the case of hypichnial preservation, which may form by casting of epigenic traces, exhumation and casting of endogenic (and intergenic) traces and collapse or active filling into an intergenic trace. As purely descriptive terms, exichnia and endichnia have advantages over the less informative full relief. Here I use, following Fillion & Pickerill 1990a and others, the terms endorelief and exorelief, parallelling the use of endichnia and exichnia in referring to position relative to casting medium. Both endoreliefs and exoreliefs are full reliefs, a term also used here.

group are found in early Cambrian Baltoscandian beds in, e.g., Scania (Hadding 1929) and Torneträsk, northern Sweden (Jensen & Grant 1992), where their disruptive activity on the sediment is considerable. *Psammichnites* is known from the Lingulid sandstone in Närke (Eklund 1961); a large form from Kinnekulle described by Högbom (1925) most probably originated from the Lingulid sandstone. Their absence from the Mickwitzia sandstone could depend on greater environmental stress as considered above. The amount of bioturbation is higher in the Lingulid sandstone than in the Mickwitzia sandstone, probably also reflecting higher infaunal activity of the macrofauna.

Systematic ichnology

Use of open nomenclature follows recommendations by Bengtson (1988).

Ichnogenus *Bergaueria* Prantl, 1946

Bergaueria perata Prantl, 1946
Fig. 24

Material. – Figured SGU 8583. Four additional specimens in collection of SGU. Lugnås. Possibly similar but larger specimens were observed about 1.5 m above the basement in the Minnesfjället mine. Eight specimens on two slabs from the Mickwitzia sandstone at Svartåfors, Östergötland, PMU Ög 138 (figured), 139. Additional specimens from Östergötland in the RM collections.

Description. – Plug-shaped, cylindrical to gently conical burrows with rounded to nearly flat base (Fig. 24). In one burrow (SGU 8583) the basal part is of smaller diameter than the main part of the burrow (Fig. 24A, B). Lower end smooth or with a very faint radiating sculpture, and in most specimens with a low central depression. One specimen has a shallow cleft dividing the base into two about equal halves. The sides of the burrows have faintly developed concentric ornamentation (Fig. 24D). The best preserved specimen (Fig. 24A, B) is 27 mm long and has a diameter at the top of about 40 mm; the base is 20 mm wide and 6 mm deep. Length/width for two other burrows are 23/40 mm and 20/41 mm. The burrows are preserved in full relief. Internal structures have been observed only in chipped specimen and reveal no signs of a marked lining. One specimen has a core of cleaner sandstone.

Discussion. – Pemberton *et al.* (1988) diagnosed *Bergaueria perata* as bergauerians with 'smooth walls; rounded lower end may exhibit faint radial ridges emanating from one or more weak central depression'. Pickerill (1989) extended the diagnosis by including concentric ornamentation as a character. The absence of a central depression in one specimen may be caused by adhering material. The lack of a central depression is essentially the one characteristic that distinguishes *Bergaueria hemispherica* Crimes *et al.* 1977 from *B. perata*. It seems that the justification for considering these as separate ichnospecies is yet to be made; a reported co-occurrence of *B. perata*, *B. hemisphaerica* and *B. radiata* in the Lower Cambrian of Alberta (Pemberton & Magwood 1990) is not convincing. The small variations exhibited by the burrows could be caused by preservational variations. Though slightly more conical than typical bergauerians, the burrows described above fall within the morphologic

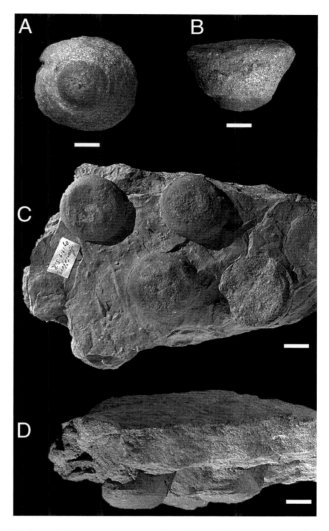

Fig. 24. □A, B. *Bergaueria perata* in lower (A) and side (B) view. Scale bar 10 mm. Lugnås. Probably interval B. SGU 8583. □C, D. *Bergaueria perata* from the Mickwitzia sandstone in Östergötland (Svartåforsen), in lower (C), and side view (D). A central hollow is faintly visible on some specimens. Scale bar 10 mm. PMU Ög 138.

range of *Bergaueria perata* as given by Pemberton *et al.* (1988). With their conical shape and smooth, rounded basal part, their strongest similarity to illustrated specimens may be to *Bergaueria major* Palij, 1976, from the early Cambrian Khmelnitsky Formation of Podolia, Ukraine (Palij 1976; Palij *et al.* 1983) and from the Mazowsze Formation of the Lublin area, Poland (Pacześna 1986). Pemberton *et al.* (1988) considered *B. major* to be synonymous with *B. perata*.

Bergaueria is generally considered to be a dwelling or resting burrow of actinarians (Pemberton *et al.* 1988).

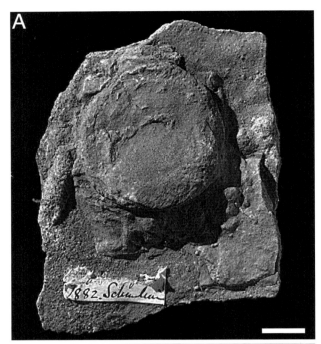

Fig. 25. □A. *Bergaueria sucta.* Preserved in hyporelief. Lugnås. Scale bar 10 mm. SGU 8584. □B. Side view of slab with possible type material of *Diplocraterion lyelli* Torell, 1870. Scale bar 10 mm. Lugnås. SGU Type 5366.

Bergaueria sucta Seilacher, 1990

Figs. 25A, 66?

Synonymy. – □?1985 *Beltanelloides simplex* (Pal.) – Gureev, Velikanov & Ivanchenko, Figs. 1v, 2g. □? *pars* 1988 *Beltanelloides simplex* (Pal.) – Gureev, Pl. 10:1. □1990 *Bergaueria sucta* n. ichnosp. – Seilacher, p. 666, Fig. 32:12, Pl. 32:2a–b.

Material. – Four slabs from the Mickwitzia sandstone at Lugnås and Billingen. Figured SGU 8584. Additional specimens in collection at RM.

Description. – Circular low-relief discs on the base of beds, commonly laterally repeated in overlapping series. May occur inclined to the base of the bed resulting in a crescentic outline. Marked marginal rim with steep vertical borders. Towards the margin there may be radial, spiralling markings. The base of the disc is gently concave. A well-preserved specimen has a diameter of 30–35 mm and a length of about 3 mm (Fig. 25A)

Discussion. – Seilacher (1990) erected this species for what was interpreted as impressions of the basal disc of an actinian cnidarian, with multiple sideways repetitions interpreted as formed during creeping (Seilacher 1990, p. 666). Fedonkin (1983, Pl. 31:6, 1985, p. 114–115) suggested a similar formation for fossils identified as *Bergaueria* sp. from the Vendian of Podolia, Ukraine. These are globular structures with lateral movement resulting in a series of meniscus pads of sediment. Probably assignable to *Bergaueria sucta* are forms reported as *Beltanelloides simplex* from the Khmelnitsky Formation of Podolia (Gureev *et al.* 1985, Gureev 1988). Illustrated specimens (Gureev 1988, Pl. 10:1) do not show the typical lateral movement of *Bergaueria sucta* but otherwise have a highly similar low flat disc.

Fedonkin & Runnegar (1992) and Runnegar (1992) have suggested that some of the Vendian forms attributed to *Nemiana* (also as *Beltanelliformis* and *Beltanelloides*) may have been formed in a similar way to *Bergaueria*. *Bergaueria sucta*, if correctly interpreted, further strengthens this possibility.

It must be stressed that *B. sucta* is easily misidentified with physical sedimentary structures.

Ichnogenus *Brooksella* Walcott, 1896

?*Brooksella* isp.

Fig. 26

Material. – Figured slab SGU 8579, from Lugnås, with four specimens; level unknown. Box label reads 'Sandsten

Fig. 26. Brooksella? isp., seen in lower (A) and lower oblique (B) view. Scale bar 10 mm. Lugnås. SGU 8579

(Medusina)'. One slab from Stora Rud, Lugnås, with two poorly preserved specimens.

Description. – Trace fossil consisting of radially arranged knob-like segments preserved in positive hyporelief (Fig. 26). The knobs are up to 7 mm long and about 3 mm in diameter, directed horizontally or gently inclined relative to bedding (Fig. 26). Width of the entire structure increases upwards (Fig. 26B), reaching a diameter of about 1.5 cm. Preservation of trace is too poor to reveal if there is an ordered vertical arrangement of the knobs. A few knobs have fine radiating markings on their tops.

Discussion. – This trace probably represents repeated feeding probes made from a common center. The radiating markings at distal parts of well-preserved probes are identical to markings seen on burrows figured by Jensen (1990, Fig. 6C), and may be impressions of a similar digging apparatus. These burrows exhibit a crude similarity to specimens of *Brooksella confusa* Walcott, 1896, from the Middle Cambrian of Alabama (e.g., Walcott 1896, Pl. 28:7). Originally interpreted as coelenterates, *Brooksella* is now mostly interpreted as complex feeding burrows (Fürsich & Bromley 1985; Seilacher 1992a), and Fürsich & Bromley (1985) put *Brooksella* in synonymy with *Dactyloidites*, consisting of vertical radial spreiten structures possessing a central shaft. The Mickwitzia specimens as well as several specimens illustrated by Walcott (1896) appear to lack a clear spreite arrange-

ment and rather consist of irregularly arranged circular lobes. Thus the name *Brooksella* could be retained for the irregular lobate forms, whereas *Dactyloidites* is used for specimens with a higher order of well-developed spreite. As seen from the work of Walcott (1896, 1898), there is great variability among these forms. A radiating spreite has been reported in Middle Cambrian *Brooksella* (Fürsich & Bromley 1985; Seilacher 1992a), but no detailed information has yet been given; pending this, the present burrows are compared to *Brooksella* rather than *Dactyloidites*. *Brooksella* includes several problematic species of doubtful metazoan connection. The Precambrian *Brooksella canyonensis* has been recognized as a trace fossil by Glaessner (1969), Kauffman & Fürsich (1983) and Fürsich & Bromley (1985), though an organic origin was questioned by Cloud (1968, 1973) and McMenamin & Schulte McMenamin (1990); Fedonkin & Runnegar (1992) consider it to be a pseudofossil. *Brooksella silurica* (Huene, 1904), from Ordovician carbonates of Sweden, has been interpreted as a medusae (Huene 1904). It has a central core consisting of a shell or skeletal debris surrounded by a roughly circular flattened body. Its interpretation is not clear, but affiliation with medusae can be excluded and its formation was probably controlled by algal mats (Ulf Sturesson, Uppsala, personal communication 1992). Probably related structures were illustrated by Holmer (1989, Fig. 4) and were suggested to represent stromatolites.

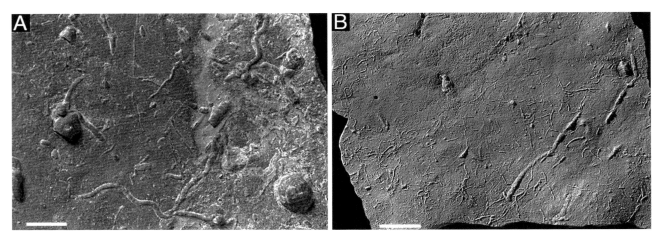

Fig. 27. □A. *Cochlichnus* isp., occurring together with small straight horizontal burrows and a *Monocraterion* cf. *tentaculatum* (lower part of picture). Preserved in hyporelief. Scale bar 5 mm . Hällekis. RM X3298. □B. *Helminthoidichnites tenuis.* Thin burrows with irregular course. Burrow crossings common, though self-level crossing are very rare. Scale bar 5 mm. Lugnås. RM X3299.

Ichnogenus *Cochlichnus* Hitchcock, 1858

Cochlichnus isp.

Fig. 27A

Material. – Two slabs with two well preserved and additional fragmentary specimens; figured RM X3298. Hällekis.

Description. – Sinuous full-relief burrows pressed onto base of sandstone slabs (Fig. 27A). Width of burrows about 0.7 mm, longest preserved segment about 2 cm. Amplitude of winding low, and course not very regular within preserved segments (Fig. 27A). Surface of burrow with a very faintly developed annulation. *Cochlichnus* isp. is found together with broken remnants of similar-sized inclined segments. The two most well-preserved specimens meet small vertical cones of greater dimension.

Discussion. – These traces are not as regularly sinuous as typical *Cochlichnus* but are broadly similar in size and appearance to *Cochlichnus serpens* Webby, 1970, from the Lower Cambrian of New South Wales. Differences between ichnospecies of *Cochlichnus* (at least five) is not clear, and McCann & Pickerill (1988) suggested that *C. serpens* (and *C. kochi* Ludwig, 1869) is conspecific with *C. anguineus* Hitchcock, 1858. *Cochlichnus* is known from the late Vendian (Fedonkin 1983) and continues throughout the Phanerozoic. It is found in various marine and non-marine environments (Hakes 1976). As producer of the trace, annelids and especially nematodes have been suggested (Seilacher 1963; Moussa 1970). Elliot (1985) found Carboniferous non-marine *Cochlichnus* from England to be full-relief tunnels, probably made by nematodes progressing along near-surface laminae with a suitable consistency for propulsion by sinusoidal waves. The Mickwitzia specimens are preserved as sand-filled tunnels made in silty–clayey sediment.

The two specimens meeting vertical cones are reminiscent of *Sokolovichnites angelicae* Gureev, 1983, from the early Cambrian Khmelnitsky formation of Podolia, Ukraine. Gureev (1983) erected *Sokolovichnites* for winding ridges in which one end gradually fades into the sediment and the other end stops at a vertical object which is wider than the ridge. This was interpreted as stimuli-guided food search with the terminal expansion probably showing place of encounter with source of stimuli, which may have been a living or dead benthic organism (Gureev 1983). Some of the illustrated specimens (Gureev 1983, Figs. 7, 8) end in structures very similar to those in the Mickwitzia specimens. Eagar *et al.* (1985, p. 129, Pl. 12F) considered an occurrence of numerous *Cochlichnus* radially arranged around a *Lockeia* in Carboniferous sediments of England to represent worms attracted by a decaying bivalve. The associations in the Mickwitzia sandstone specimens may be accidental.

Ichnogenus *Cruziana* d'Orbigny, 1842

Most authors use *Cruziana* and *Rusophycus* as separate ichnogenera, emphasizing the predominant horizontal expression of the former and vertical expression of the latter (e.g., Osgood 1970; Crimes 1970; Birkenmajer & Bruton 1971; Bromley 1990). In an ethological classification, *Cruziana* and *Rusophycus* are commonly classified as, respectively, repichnion and cubichnion, though the motivation behind the traces may be the same. As pointed out by Bergström (1976), most *Cruziana* are too deeply impressed to be purely locomotory structures, and the primary purpose behind both structures were more likely feeding; microphagous (Schmalfuss 1981; Seilacher 1985) and/or macrophagous (Bergström 1973a; Jensen 1990). In the case of *Rusophycus*, resting, dwelling and protection could also be considered likely activities (e.g., Osgood 1970).

End members of *Rusophycus* and *Cruziana* are morphologically distinct, but there are intermediate forms for which assignment is difficult. Intermediate forms seem to be especially common in the lower Palaeozoic, a situation well illustrated by early Cambrian specimens from Poland that were provisionally named *Cruziana rusoformis* and *Rusophycus cruziformis* by Orłowski *et al.* (1970). Osgood (1970) addressed the problem of repeated *Rusophycus* by stating that true *Cruziana* have an even lower surface, while horizontally repeated *Rusophycus* reveal the alternation of horizontal and vertical digging direction. However, this is true only for repetition of strongly convex *Rusophycus*; the smoothness of horizontally repeated flat *Rusophycus* would depend on how closely the individual burrows overlap. Examples of this overlap are visible also in such 'typical' cruzianiform species as *Cruziana semiplicata* Salter, 1853 (see e.g., Pl. 4 of Orłowski *et al.* 1970). Little consideration seems to have been given to whether, or how commonly, *Cruziana* consists of horizontally repetitive *Rusophycus*. *Cruziana* is generally thought to have been made in continuous horizontal furrowing (Crimes 1970; Birkenmajer & Bruton 1971), the propulsion being achieved by a forward-progressing wave of leg movements (Seilacher 1970). In several species of *Cruziana* the angle at which the scratch marks meet varies within the length of the trace, often with overlapping sets (e.g., Birkenmajer & Bruton 1971, Fig. 10). This has been explained by alternating force exerted by the legs in a backward-proceeding wave of leg movement. However, the notion of continuous horizontal propulsion is suspect. First, the angle at which the scratches meet medially is in several species of *Cruziana* very high, and as the scratches are more or less straight, the resulting net forward movement would be low. As pointed out by Whittington (1980) the tip of an arthropod leg is kept in one spot during propulsion. For the same reason also specimens with scratches of low V-angle and higher curvature give no definite indication of forward propulsion; after all, the same features are seen also in decidedly stationary *Rusophycus* burrows. Most *Cruziana* have been ascribed to trilobites. Whittington (1980) found it impossible to reconcile *Cruziana*-making activity with the known leg morphology of *Olenoides serratus* and also considered it unlikely for trilobites as a whole. The few trilobites with known leg morphology need not be representative, and *Cruziana* may certainly have been produced by arthropods other than trilobites,

Whittington's (1980) work should inspire great caution in the interpretation of *Cruziana*.

None of the specimens of *Cruziana* found in the Mickwitzia sandstone show evidence of being formed in continuous horizontal motion. This is most clearly seen in *Cruziana tenella* were the surface often is highly undulatory, but the deepest part of each burrow is of about the same depth and forms a continuous chain. This leads to the question of how closely burrows of *Rusophycus* should overlap to be a *Cruziana*. Here, *Cruziana* is used for horizontally repetitive *Rusophycus*, not necessarily overlapping, arranged in a more or less straight line and dug to about the same depth. In several specimens where a *Cruziana* ends with a *Rusophycus*, the latter is more deeply dug into the sediment, probably reflecting different motivation of the two burrows (Fig. 29). Problems arise with burrows that were not merely added at a more or less constant depth. An example of this is seen in burrows made in location of prey (Bergström 1973a; Jensen 1990), some of which are flatly U-shaped in profile (Fig. 30). These were identified as *Rusophycus dispar* by Jensen (1990), but their repetitive nature, though not strictly horizontal, makes assignment to *Cruziana* more proper.

Ichnospecies of *Rusophycus* and *Cruziana* are to a large extent based on morphologic features resulting from leg morphology of the producer, of which they give information to a degree not common in trace fossils (Seilacher 1970). As these traces are often assigned to trilobites, the body fossils of which are useful stratigraphic tools, their potential for stratigraphic use in beds otherwise lacking diagnostic fossils has been investigated (Crimes 1968, 1970, 1975a; Crimes & Marcos 1976; Seilacher 1970, 1990, 1992b; Seilacher & Crimes 1969). Though there are instances of apparent stratigraphic usefulness, the repeated development of similar claw patterns is still poorly documented (Magwood & Pemberton 1990). A matter of some complication is that in the sole monographic work on the subject (Seilacher 1970) *Cruziana* and *Rusophycus* were synonymized. This work introduced several new species, some of which have subsequently been divided into species of *Rusophycus* and *Cruziana* by later authors (Crimes *et al.* 1977; Legg 1985).

Cruziana rusoformis Orłowski, 1992

Figs. 28B, 29, 30

Synonymy. – □*pars* 1869 *Rusophycus dispar* n.sp. – Linnarsson, pp. 353–356. □*pars* 1870 *Cruziana dispar* Lins. sp. – Torell, p. 6. □*pars* 1871 *Cruziana dispar* Linnarsson – Linnarsson, pp. 14–16, Fig. 19, *non* Figs. 17, 18 (= *Rusophycus dispar*). □1901 *Cruziana* – Holm, Fig. 3. □1970 *Rusophycus* sp. – Banks, Pl. 3c. □*pars* 1990 *Rusophycus dispar* Linnarsson, 1869 – Jensen, p. 31. Figs. 2, 3,

Fig. 28. □A *Cruziana* cf. *rusoformis*, with transvere coarse scratch marks and somewhat finer longitudinally directed scratch marks. Scale bar 10 mm. Interval D, about 8 m above basement, about 600 m south of Hälle-kis. RM X3300. □B. *Cruziana rusoformis.* Scale bar 10 mm. Hjälmsäter. Interval B. RM X3301.

4b, 8, *non* Figs. 1, 4a, 9, 10 (= *Rusophycus dispar* Linnarsson, 1869). □*pars* 1970 '*Cruziana rusoformis* ichnosp. nov.' – Orłowski, Radwański & Roniewicz, p. 356, Pl. 1b, d, *non* Pls. 1a (= *Rusophycus dispar* Linnarsson, 1869), 1c (= *Rusophycus radwanskii* Alpert, 1976), 2b (= *Cruziana regularis* Orłowski, 1992). □1974 *Cruziana rusoformis* Orłowski, Radwański & Roniewicz, 1970 – Orłowski, p. 7, Pl. 6:1, 2. □1992 *Cruziana rusoformis* Orłowski, Radwań-ski and Roniewicz, 1970 – Orłowski, pp. 19, 23, Fig. 5:1– 3. □*pars* 1992 *Cruziana dispar* Linnarsson, 1871 – Orłowski, p. 18, Fig. 4:1–4, *non* Fig. 3.

Material. – *Cruziana rusoformis* has been found in Lugnås, Kinnekulle and Billingen. Common in interval B. Figured specimens SGU 8585, RM X42, 3301. Specimens in collection at RM and SGU.

Description. – Straight or gently curved traces with two parallel lobes, each possessing sharp ridges, separated by a median furrow (Figs. 28B, 29). The two lobes may be in contact or, in shallow forms, be separated by a medial flat area (Fig. 29). The ridges are straight or, more commonly, somewhat curved, and meet to form an angle of 120°– 180° degrees, only rarely less. The angle varies within the length of a trace; in short specimens often with a gradual forward increase in V-angle, generally corresponding with an increased width of the trace. The scratches often occur in sets of two but also sets of three and possibly more (Fig. 28B). Width and depth varies within a trace; short forms may have a gently U-shaped profile, and con-tinuations into typical *Rusophycus* are not rare. The width of the trace ranges from 13 to 65 mm; length is up to about 250 mm.

Discussion. – In his original descriptions of *Rusophycus dispar*, Linnarsson (1869a) included both rusophyciform and cruzianiform specimens and noted that every transi-tion between the two exists. For reasons of original desig-nation, *R. dispar* is here restricted to rusophyciform bur-rows. Very similar burrows occur in the Lower Cambrian of Poland. These burrows were presented under informal taxonomy as '*Cruziana rusoformis* ichnosp. nov.' by Or-łowski *et al.* (1970, Pl. 1b, d, 2a), who explicitly stated (p. 356) that *C. rusoformis* was not introduced as a formal taxon. This was motivated by anticipations that the trace fossil conference in Liverpool, 1970, should bring a clear consensus on the characters on which ichnotaxonomy should be based (Orłowski *et al.* 1970, p. 356, 1971, pp. 343–345). Orłowski (1974) gave some additional com-ments on *C. rusoformis* concerning its assumed mode of formation. However, no formal designation was made, and under article 15 of the International Code of Zoolog-ical Nomenclature, *C. rusoformis* Orłowski, Radwański & Roniewicz, 1970, is not an available taxon. The first for-mal treatment of *C. rusoformis* is that of Orłowski (1992). Seilacher (1970) and Bergström & Peel (1988) thought it

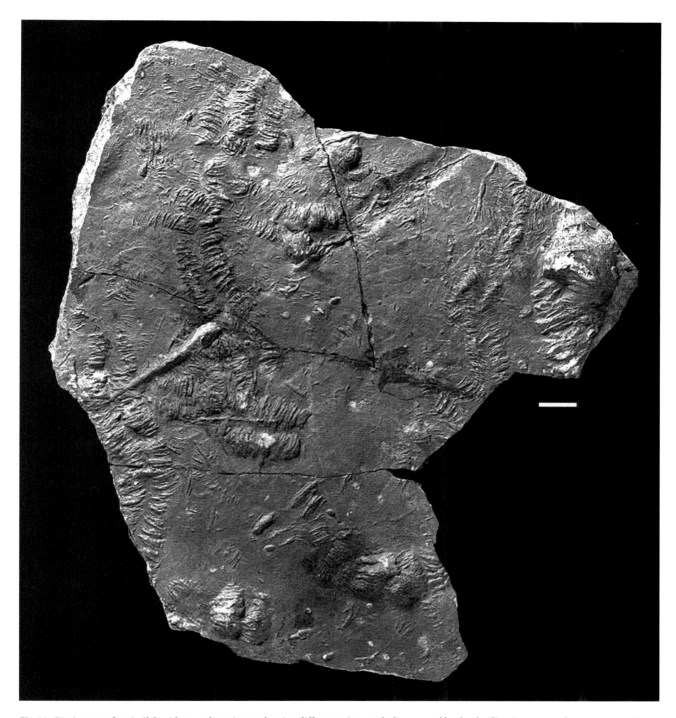

Fig. 29. Cruziana rusoformis. Slab with several specimens showing differences in morphology caused by depth of impingement. Also seen are specimens of *Rusophycus dispar*, including some that are continuous with *Cruziana rusoformis.* Scale bar 20 mm. Interval B. Lugnås. RM X42.

likely that *C. rusoformis* is a synonym of *C. dispar*, a view shared here. Orłowski (1992) maintained the two as separate species, stating a difference in size, which is not evident from the measurements given, and a difference in claw formula. This difference in claw formula is not well founded in Orłowski's description. *C. rusoformis* is said to have 'fine scratches near the median furrow, on lateral sides associated with bunches of three delicate, sharp scratches' (Orłowski 1992, p. 19), whereas the claw formula of *C. dispar* is not specified, though given as different, and it is also stated that the difference is 'clearly visible in large collections'. From the discussion above, and

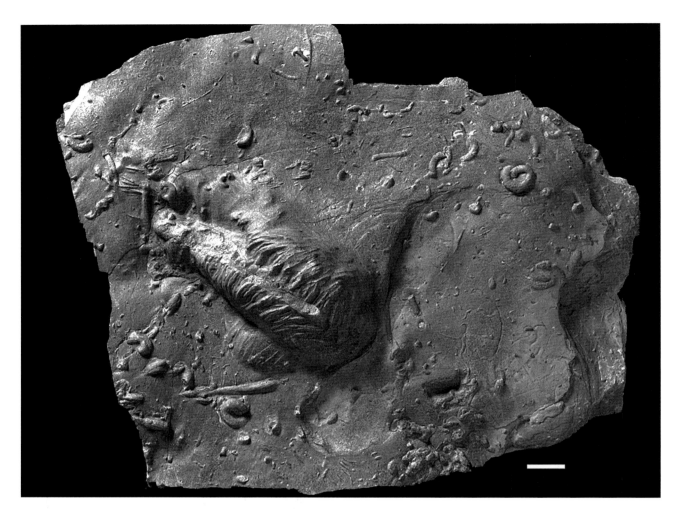

Fig. 30. Association of *Cruziana rusoformis* with worm burrow interpreted to represent hunting behavior of a trilobite. Also seen are specimens of *Gyrolithes polonicus*. Scale bar 20 mm. Lugnås. SGU 8585.

judging from the illustrated specimens, the two are considered synonymous. *C. rusoformis* also occurs in the Norretorp Sandstone in Scania (personal observation, 1990). Another similar form is *Cruziana salomonis* Seilacher, 1990, from the Lower Cambrian of Jordan and Sinai. According to Seilacher (1990, p. 662), *C. salomonis* has four secondary claws, which is more than what is typically observed in *C. rusoformis*. However, considering that the claw pattern in *Rusophycus dispar* (a trace that at least in some cases had the same producer) includes forms with four secondary claws, this distinction would rather indicate a different digging attitude. Judged from specimens illustrated by Seilacher (1990, Pl. 32.2c–g), sets of claw-impressions are more distinctly seen in *C. salomonis* than in *C. rusoformis*. Another large form with high V-angle, *Cruziana warrisi* Webby, 1983, from Lower Ordovician of New South Wales, differs by having finer and denser scratch marks with more constant V-angle (Webby 1983). An interesting comparison may be made with

Cruziana canonensis Fischer, 1978, from the Ordovician of Colorado. This large form, about 10 cm wide, consists of 8–10 pairs of lobes, with longitudinal markings indicating five-clawed legs. The last two lobes curve posteriorly and become parallel to the body axis. Anterior to this, the lobes are directed with a V-angle of 120°–130°, with the first three lobes being shallow and the next five deep (Fischer 1978, p. 24). Fischer (1978) believed the limb motion to have been opposite to that inferred for trilobites and suggested an aglaspidid, which occurs in the same formation as the producer. However, the reason for the supposed outward–posterior leg movement was not given. As pointed out by Fischer (1978, p. 24), there are general similarities between *C. rusoformis* and *C. canonensis*. The question of synonymy must be left open, but the five-clawed pattern in *C. canonensis* seems distinct.

Some *Cruziana rusoformis* were dug mainly within sand, with relatively little impingement into the underlying clay. This is most clearly seen in shallow burrows in which the width increases with depth (Fig. 29). The motivation of *C. rusoformis* has in some specimens been shown to be search for infaunal soft-bodied animals (Bergström 1973a; Jensen 1990). Others may have been made during deposit feeding, as depicted by Seilacher (1985), though the means by which this was accomplished remains problematic.

Distorted specimens of *C. rusoformis* have been found in large, elongated crudely bilobate structures with blunt ridges. The few specimens of this type are all more than 10 cm wide, which makes them wider than any well-preserved specimen of *C. rusoformis*, though within the range of *Rusophycus dispar*. Along parts of some specimens, sharper scratch marks are still visible. The cause of deformation was probably scouring currents, possibly combined with some soft-sediment deformation. A specimen of this type preserving only one side of distorted scratch marks was illustrated by Eklund (1961, Fig. 13). Within the distorted lobe that specimen has two shells of *Mickwitzia monilifera*, further indicating that the trace has been exhumed.

Cruziana cf. *rusoformis* Seilacher, 1970

Fig. 28A

Material. – One specimen, RM X3300. Found about 600 m south of Hällekis in a small outcrop with material in scree at the shore of Vänern, implying a level higher than 8 m above the basal gneiss. Preserved as isolated gutter of medium-to-coarse-grained sandstone.

Description. – Fragmentary specimen consisting of inner area with coarse oblique scratch marks bordered by an outer area with somewhat finer, longitudinally directed scratches (Fig. 28A). Inner scratches are straight to gently curved, with convex side facing forward, occurring in pairs, possibly in groups of three, meeting medially at an angle of 130°–150°. The coarse scratches are grouped in two longitudinally overlapping sets; the posterior and best preserved have a posteriorly tapering width from about 55 mm to 25 mm (Fig. 28A) The widest part is also the deepest part of the trace, reaching about 20 mm. The outer, finer scratches form 35 mm wide borders in the posteriormost part of trace, decreasing in width anteriorly and possibly entirely missing in the deepest part. Scratches occur in groups of at least six, with distance between outer pairs in one prominent group being 6.5 mm. On the extreme sides of the burrow are more prominent, though poorly preserved ridges.

Discussion. – This trace has longitudinal scratches in addition to transverse ones. Such scratches are generally thought to be made by the outer leg-branch, whereas transversely directed scratches are assigned to the inner leg-branch (Seilacher 1970; Bergström 1973a, 1976). Comparison on species level of this specimen is difficult because of fragmentary preservation. The transverse scratches are identical to those on *C. rusoformis*, and it may therefore represent a rare preservation of the ichnospecies. Rare exite brushings have been reported from the closely related *C. barbata* Seilacher, 1970, and *C. salomonis* Seilacher, 1990 (Seilacher 1990, 1992b). Some similarity exists also with *C. arizonensis* Seilacher, 1970, from the Middle Cambrian Flathead Sandstone in Montana (Walcott 1918, Pl. 39:3, 4; Seilacher 1970). The main difference appears to be that the outer border in the Flathead Sandstone specimen is a smooth brushed surface, whereas in this specimen there are distinct scratches. The zone with longitudinally directed scratches seems to have reached to the center of the trace, though they are largely concealed by the deeper transverse scratches. Cross-cutting relationships show that the transverse scratches postdate the longitudinal ones. Formation of the longitudinally directed scratches is problematic. Their spacing, size and numbers within sets are comparable to the scratches on the anterior side of the proverse diggings in *Rusophycus dispar*. If there were hinge joints between podomeres and a mainly transverse axis of swing between coxa and its site of attachment (cf. Whittington 1980), a strong backward rotation along the axis of swing and strong flexure of the podomeres would bring the anterior side in contact with the sediment. This would imply a stance with the supporting legs held out flatly and probably with the digging carried out by only a few pairs of legs. This mode of digging could have been used to stir up sediment towards the mouth. This interpretation is an alternative to the common explanation of similar structures as formed by raking exites, an interpretation that has been found functionally unlikely in the two well-studied trilobites *Olenoides serratus* (Whittington 1975, 1980) and *Triarthrus eatoni* (Whittington & Almond 1987). The prominent lateral ridge was probably made by the lateral parts of the exoskeleton.

Whether this trace fossil represents an unusual behaviour or the activity of a different type of olenellid trilobite, must be left open. Furthermore, though a trilobite is a highly likely producer because of similarity of the transverse scratch marks in *Rusophycus dispar*, other types of arthropods cannot be ruled out. The case for an olenellid producer is strengthened by a *Rusophycus* burrow from the lower part of the Lükati Formation in the Kopli quarry (level according to K. Mens, personal communication, 1992), Tallinn, Estonia, with fine longitudinal as well as transverse scratches (personal observation, 1992). In size, the Estonian specimen matches *Schmidtiellus reetae* Bergström, 1973, known from a comparable level, the *Volborthella* Zone, in eastern Estonia (Bergström 1973b, p. 301ff).

Fig. 31. Lower surface of slab with abundant *Cruziana tenella* and coffee-bean-shaped *Rusophycus eutendorfensis.* Scale bar 5 mm. Lugnås. SGU 8586.

Cruziana tenella (Linnarsson, 1871)

Figs. 31, 32

Synonymy. – □1871 *Fraena tenella* n.sp. – Linnarsson, p. 11, Pl. 1:5. □*pars*1921 *Ichnium problematicum* – Schindewolf, pp. 36–40. For further synonymy, see *Cruziana problematica* (Schindewolf, 1921) *in* Bromley & Asgaard (1979), pp. 66–67.

Diagnosis. – Small *Cruziana* having transverse to nearly transverse striae. Tendency for grouping of striae in pairs in some specimens. Observed range of width 0.8–11 mm. (Based on Bromley & Asgaard 1979, p. 68.)

Material. – Figured slabs, SGU 8586, RM X3302, 3303. Additional specimens in the collections of RM and SGU. *Cruziana tenella* is common in intervals B and D at Kinnekulle and was found at Billingen and Lugnås.

The specimens in Fig. 31 (SGU 8586) are similar to specimens illustrated by Linnarsson (Pl. 1:5). A label accompanying the slab reads 'Harlania tenella Linrs., Vg. Lugnås, G. Linnarsson'.

Description. – Small, parallel-sided traces with a central groove dividing the trace longitudinally into two lobes (Figs. 31, 32). Length of trace from a few to more than 50 millimetres; burrow mostly straight or gently winding, occasionally highly winding including crossings of older path. Commonly occurring in great numbers, at points of crossings generally passing above or below older path (Fig. 31). Sides of burrows steep, commonly vertical. Surface even to undulating, in the latter case with nodes 0.5–1 mm apart (Figs. 31, 32A). Occasionally the trace consists of a series of separated, highly arched bilobes, thereby integrating with *Rusophycus eutendorfensis* (Fig. 32B). Rare-

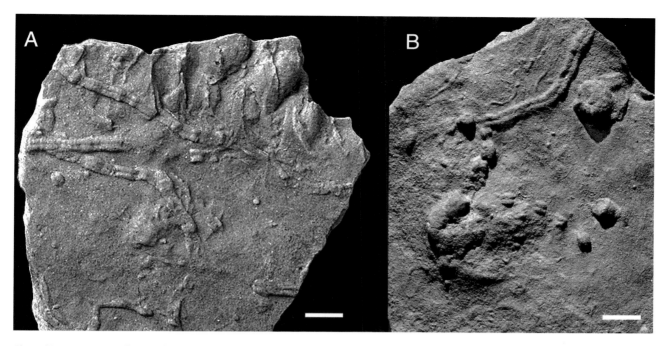

Fig. 32. □A. *Cruziana tenella.* Note fine transverse ornamentation. Scale bar 5 mm. Hjälmsäter. RM X3302. □B. *Cruziana tenella* and *Rusophycus eutendorfensis.* The latter consists of isolated coffee-bean-shaped traces. Scale bar 5 mm. RM X3303.

ly the lobes preserve delicate striae at about right angle to trace extension (Fig. 32A). Width of trace 0.8–2.2 mm; depth 0.4–0.9 mm.

Discussion. – These traces were described as *Fraena tenella* by Linnarsson (1871) who brought the species to *Fraena* Rouault, 1851, with some reservation due to the heterogenic nature of that genus (there is an English translation of Linnarssons discussion on *Fraena* in Matthew 1891, pp. 158–159). Following the recommendation of Tromelin & Lebesconte (1876; see also Péneau 1946), *Fraena* is today restricted to the unilobate type species *Fraena sainthilairci* Rouault, 1851, which is similar to *Arthrophycus* (Durand 1985a).

According to Nathorst (1881b), *Psammichnites impressus* Torell, 1870, and *Psammichnites filiformis* Torell, 1870, are identical to *Fraena tenella*. Torell's (1870) description of *Psammichnites impressus* agrees well with the morphology of *Fraena tenella*. However, an illustration of the same fossil given as 'Petrefactum incertae sedis' by Torell (1868, Pl. 3:4) shows a trace which appears to be trilobate as well as bilobate. As no type material of *P. impressus* have been found, its relation to the bilobate trace is uncertain. Lacking type material and with the above mentioned inconsistencies and uncertainties, *P. impressus* is considered a *nomen dubium*. The description of *Psammichnites filiformis* does not state whether the trace is uni- or bilobate. Examination of type material of

P. filiformis shows this to be winding, apparently cylindrical tunnels considerably different from *F. tenella*.

Linnarsson (1871) pointed out that *Fraena tenella* is similar to *Arthrophycus harlani* in possessing a longitudinal furrow and transverse divisions and in the shape of its cross-section, and that the main difference is in size. *Arthrophycus* is mostly strongly bundled (Häntzschel 1975) and similar to *Phycodes* in its organization. This differs from the more random orientation of *Fraena tenella*. In their morphology and in their occurrence together with, and gradation into short rusophyciform burrows, these traces are identical to forms commonly referred to as *Isopodichnus* Bornemann, 1889, and Seilacher (1960, Pl. 1:4) illustrated Mickwitzia sandstone specimens for comparison with *Isopodichnus*. *Isopodichnus* as used by most authors is bimorphic, consisting of elongated cruzianiform, and short rusophyciform elements. They may be seen as small specimens of *Cruziana* and *Rusophycus*, respectively, and the problems of differentiating *Isopodichnus* from *Cruziana* and *Rusophycus* have been repeatedly addressed (see Walter 1984; Pollard 1985; Bromley 1990). Often *Isopodichnus* have been used for freshwater traces supposedly made by small crustaceans, but no good criteria have been found to distinguish them from similar marine traces. In view of this and the probable ethological similarity, I follow the procedure of Bromley & Asgaard (1979) and Bromley (1990) in considering *Isopodichnus* as superfluous, assigning the elongated forms to *Cruziana* as defined in the discussion on *Cruziana rusoformis*. Though *Fraena tenella* Linnarsson, 1871, is a rather obscure name, which for reasons of stability could be rejected in favor of *Cruziana problematica* (Schindewolf, 1921), it is retained to rid this type of trace from the environmental biases coupled to the species *Ichnium problematicum*. The problem inherent in using size as a main identifier for *Cruziana tenella* – 'size alone is almost sufficient to distinguish it from other species' Bromley & Asgaard (1979, p. 68) – is less severe at the ichnospecific than ichnogeneric level.

A serious problem, however, could be to distinguish it from traces usually interpreted to be made by animals with a molluscan-type foot; most notably *Didymaulichnus* Young, 1972. Most reports of *Didymaulichnus* give the surface of the lobes as smooth, sometimes undulating but without scratch marks (e.g., Hakes 1976). However, specimens possessing 'scratch marks' have also been included, and Devonian specimens from Antarctica given as *D. lyelli* and *D. nankervisi* Bradshaw, 1981, bear markings thought to be made by the digging activity of arthropods (Bradshaw 1981). It is recommended that *Didymaulichnus* is used only in its original sense (Young 1972) of burrows with smooth surface. The possibility that the smoothness is caused by preservation is a major problem demanding special care. For example, *Rouaulti rouaulti* (Lebesconte, 1883) of Crimes (1970), being a preservational variety of *Cruziana semiplicata*, should clearly be excluded from *Didymaulichnus*. Typically, the width of *Didymaulichnus* is in the range of 10–20 mm, but specimens as small as a few millimetres were reported by Crimes & Anderson (1985), making preservational aspects all the more important. *D. miettensis* Young, 1972, can be distinguished by possessing lateral bevels (Young 1972; Walter *et al.* 1989), though these may be absent where the trace is shallow.

Beside occasional transverse striae, which are interpreted as scratch marks, the main rational for an arthropodan origin of the Mickwitzia *Cruziana tenella*, is their association with and gradation into rusophyciform traces, in a manner very similar to that in Triassic forms (e.g., Schindewolf 1928, Fig. 6). Other Lower Palaeozoic occurrences of probable *C. tenella* include: Upper Cambrian of Northern Argentina (Manca 1986, Pl. 2:1), Upper Cambrian to Lower Ordovician of Newfoundland (Bergström 1976, Fig. 18; Fillion & Pickerill 1990a) and Lower Cambrian of California (Alpert 1976b; cruzianiform specimens of *Rusophycus didymus*).

Mesozoic occurrences of *Cruziana tenella* have commonly been assigned to the activity of notostracan branchiopods (e.g., Bromley & Asgaard 1979; Pollard 1985). Producers of Palaeozoic forms are unknown but could include small and juvenile trilobites (Seilacher 1953). Traces passing above and below older paths without significant disturbance points to an endogenic formation, indicating search for food rather than a purely locomotory purpose.

Ichnogenus *Diplocraterion* Torell, 1870

Diplocraterion parallelum Torell, 1870

Figs. 7A, 10A, 33A

Synonymy. – □1870 *Diplocraterion parallelum* n.sp. – Torell, p. 13. □1931 *Diplocraterion parallelum* Torell – Westergård, pp. 4–9, Pl. 1:1, 2, 4, ?3; Pl. 2; Pl. 3; Pl. 10:2–3.

Material. – About 20 specimen examined in detail on seven slabs; figured RM X3204. Additional specimens observed in the field. *Diplocraterion parallelum* is found in fine- to medium-grained sandstone beds throughout the Mickwitzia sandstone.

Description. – Endogenic burrows visible as slits in negative hypo- or epirelief (Fig. 10A) or as full burrows on vertical faces, consisting of a vertical spreite flanked by a U-shaped marginal tube (Figs. 17A, 33A). Material in spreite more clayey compared to surrounding sediment. Spreite protrusive with a basal position of the causative burrow. Spreite may be parallel-sided, have a gentle downward taper, or be somewhat globular.

Discussion. – *Diplocraterion* is generally interpreted as the dwelling structure of a filter-feeder animal, formed in a high-energy environment (e.g., Fürsich 1974a; Cornish 1986). Numerous reports have been made from sediments interpreted as of tidal origin (e.g., Bjertstedt & Erickson 1989 and references therein), and several studies have focused on ecological and sedimentological implications of *Diplocraterion* (Goldring 1962; Cornish 1986; Fürsich 1974a; Bjertstedt & Erickson 1989). Bromley & Hanken (1991) reported early Cambrian *Diplocraterion parallelum* from northern Norway for which a downwards expanding protrusive spreite was inferred to result from growth of the animal, and for which the consistent protrusive development was taken as a primitive character.

Diplocraterion parallelum(?) Torell, 1870, narrow form

Figs. 25B, 33B

Synonymy. – □1870 *Diplocraterion Lyelli* n.sp. – Torell, pp. 13–14. □1931 *Diplocraterion lyelli* Torell – Westergård, pp. 9–11. Pl. 4. □1973 *Diplocraterion lyelli* Torell 1870 – Bruun-Petersen, pp. 519–520, Fig. 4C.

Material. – Two slabs from Lugnås, described by Westergård (1931a) (not seen). One slab in collection of SGU, SGU 5366, labelled as possible type material One specimen observed in field on block at Hällekis originating from interval E.

Fig. 33. □A. *Diplocraterion parallelum* with partly preserved wide marginal tube, and impression of protrusive spreite. Scale bar 10 mm. Stora Stolan, Billingen. RM X3204. □B. *Diplocraterion parallelum* (?). Narrow U-shaped tube displaying no spreite. This type of burrow may correspond to *Diplocraterion lyelli* Torell, 1870. Field photograph of loose block at shore of Vänern about 160 m south of Hällekis. Length of burrow is about 80 mm. Originating from the lower part of interval E.

Description. – Vertical U-shaped burrow with gently upward-diverging arms, expanding into funnel-shaped tops. Basal connection between vertical limbs diffuse. Area between limbs with a higher content of muddy material than surrounding sediment, but no distinctive spreite visible (Figs. 25B, 33B). Topped by funnels that may have fine vertical ornamentation. Dimension of tubes 2–3 mm, distance between tubes 7–13 mm at the base. Diameter of funnel tops about 15 mm. Depth 6–8 cm. (Partly based on Westergård 1931a.)

Discussion. – The whereabouts of Torell's type material is not known, but Westergård (1931a) made a convincing case for material in the collection of the Swedish Geological Survey being identical, or very similar to the type material of *Diplocraterion lyelli* Torell, 1870. Torell (1870) gives the locality as Lugnås, but no additional specimens meeting the above description have been found in that area. The specimens were reported to occur in a 'yellowish–grayish sandstone which is less hard than the typical Mickwitzia sandstone: it agrees with layers embedded in the latter but also with the bottom stratum of the Lingulid Sandstone' (Westergård 1931a, p. 4). No spreite was observed, but its presence was inferred from a petrographic difference of the surrounding rock to that between the limbs. The specimens illustrated by Westergård (1931a) could not be found in the collection of SGU, but there is a slab (SGU Type 5366), that is labelled as possibly belonging to Torell's type material (Fig. 25B). Another specimen meeting the above description was observed on a block of medium-grained sandstone from interval E, supporting the occurrence in what Westergård called the bottom stratum of the Lingulid sandstone. *Diplocraterion lyelli* differs from *Diplocraterion parallelum* in having more closely set vertical limbs and in having a very diffuse spreite, or possibly altogether lacking one. These features make *Diplocraterion lyelli* similar to *Diplocraterion habichi* (Lisson, 1904). As a further characteristic of *Diplocraterion habichi*, Fürsich (1974a) listed upward divergence of the U-tube. *Diplocraterion lyelli* has a gradual upward divergence, not the steep flaring seen in the uppermost part of *Diplocraterion habichi*. Heinberg & Birkelund (1984) and Dam (1990a, b) assigned to *Diplocraterion habichi* long U-tubes without upward flaring, apparently basing the identification on the small distance between the tubes and the poorly developed spreite. A poorly developed spreite seems to be typical of *Diplocraterion habichi*; reported as lacking in Jurassic specimens from Portugal (Fürsich 1981) and Jurassic specimens of East Greenland (Dam 1990a). Diffuse spreiten were also reported in Cambrian specimens from Texas (Cornish 1986). Fürsich (1974a) synonymized *Diplocraterion lyelli* with *Diplocraterion parallelum*, but as discussed above there are some differences. Because of the more or less parallel arrangement of the limbs, these burrows are assigned to *Diplocraterion parallelum*, though with some doubt.

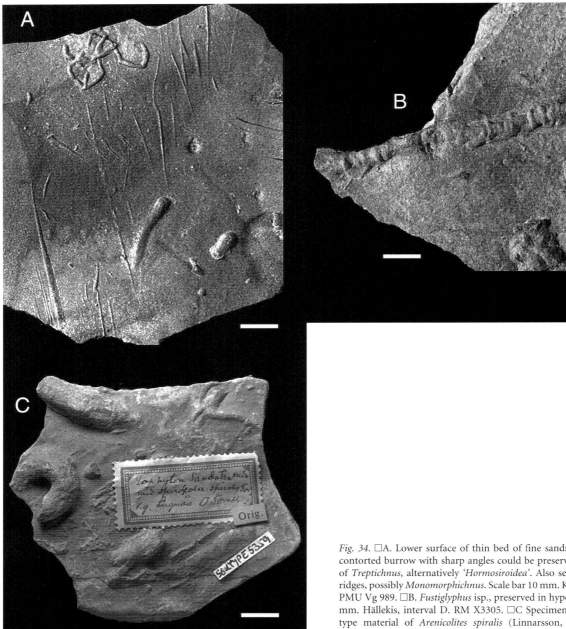

Fig. 34. □A. Lower surface of thin bed of fine sandstone. Undulating contorted burrow with sharp angles could be preservational variations of *Treptichnus*, alternatively 'Hormosiroidea'. Also seen are long sharp ridges, possibly *Monomorphichnus*. Scale bar 10 mm. Kulpetorp, Lugnås. PMU Vg 989. □B. *Fustiglyphus* isp., preserved in hyporelief. Scale bar 5 mm. Hällekis, interval D. RM X3305. □C Specimen belonging to the type material of *Arenicolites spiralis* (Linnarsson, 1869). Scale bar 10 mm. Lugnås. SGU Type 5359.

Ichnogenus *Fustiglyphus* Vialov, 1971

Fustiglyphus isp.

Fig. 34B

Material. – One specimen, RM X3305, from about 9.5 m above basement at Hällekis. Another specimen of *Fustiglyphus* isp. was illustrated by Möller (1987, Fig. 19) as 'Wurmspur mit Querringelung'. The latter specimen was collected at Hjälmsäter at 1.8 m above basement (Möller 1987, p. 46).

Description. – Gently curved burrow, about 2 mm wide, preserved as a cylindrical half-relief at base of thin bed of fine sandstone–siltstone. At irregular spacings the burrow has ring-like swellings 0.5–1.0 mm broad and about 3 mm wide (Fig. 34B). The rings appear, as far as preservation permits, to be symmetric.

Discussion. – *Fustiglyphus* consists of 'straight strings or narrow cylinders of varying length encircled by ringlike 'knots' or well-defined swellings at regular or varying intervals' (Häntzschel 1975, p. W64; Stanley & Pickerill 1993). The specimens from the Mickwitzia sandstone are comparable to the type species, *F. annulatus* Vialov, 1971, differing mainly by having less pronounced rings.

Osgood (1970) suggested that the rings represented peristaltic expansions in connection with defecation or localized muscle knotting, while Miller & Rehmer (1982, p. 891) observed that 'constriction of any fluid-filled cavity could have an effect similar to muscle-knotting'.

Suggestions that the swellings represent brood chambers (Stanley & Pickerill 1993) does not seem likely for the Mickwitzia sandstone specimens

Ichnogenus *Gyrolithes* de Saporta, 1884

Synonymy. – □? 1870 *Spiroscolex* n.g. – Torell, p. 12. □? 1892 *Daimonelix* – Barbour, p. 99, Figs. 1–3. □ 1969 *Conospiron* Vialov gen.n. – Vialov, p. 106. □ 1980 *Megagyrolithes* n.gen. – Gaillard, p. 465, Pl. 1:1–3, 7, 8.□ 1983 *Ichnogyrus*, ichnogenus nov. – Bown & Kraus, p. 111. For further synonymy see Häntzschel (1975, p. W65).

Diagnosis. – Burrows more or less describing a dextral or sinistral circular helix essentially upright in the sediment; surface with or without wall structure or scratch traces; terminally with or without expanded segment; radius of whorls and diameter of tunnel rather constant; may branch and interconnect with other burrows, often *Thalassinoides* or *Ophiomorpha* networks. (Based on Bromley & Frey 1974, pp. 317–318.)

Gyrolithes polonicus Fedonkin, 1981

Figs. 30, 34C, 35, 36C–D, 64A–B

Material. – Figured specimens on slabs, PMU Vg 990, RM X3225, 3228, 3306, SGU 8585, 8587–92. *Gyrolithes polonicus* has been found in Level B at Kinnekulle, Lugnås and Billingen.

Synonymy. – □1869 *Arenicolites spiralis* Torell – Linnarsson, p. 8. □1870 *Spiroscolex spiralis* n.g. – Torell, p. 12. □1871 *Arenicolites spiralis* Torell – Linnarsson, p. 10, Pl. 1:14. □1881 *Spiroscolex spiralis* Torell – Nathorst, pp. 28–29, Pl. 6:4. □? 1972 'Vertical, spiral traces' – Lendzion, Pl. 13:3. □? 1976 *Gyrolithes* – Garcia-Ramos, pp. 146–147, Pl. 1C. □1977 ?*Gyrolithes* sp. – Fedonkin p. 185, Pl. 5a, g. □1981 *Gyrolithes polonicus* – Fedonkin p. 80, Pl. 22, 1–5, 8. □1983 *Gyrolithes polonicus* – Fedonkin, p. 134. □?1984 *Gyrolithes* sp. – Liñan p. 57, Pl. 2:2–4. □? 1985 *Gyrolithes polonicus* – Crimes & Anderson, p. 321, Fig. 6:7, 8. □1985 *Gyrolithes polonicus* – Paczesna p. 259, Pl. 3:3. □1986 *Gyrolithes polonicus* – Paczesna p. 33, Pl. 3:2, 3

Diagnosis. – Sinistral or dextral vertical spiral burrows, with diameter of whorl up to 40 mm and diameter of burrow 1–15 mm, but mostly less than 8 mm. Surface of burrow smooth or with transverse ridges. Mostly occurring as sandy–silty filled burrows in silty–clayey sediment.

Description. – Vertical spiral burrows commonly occurring as silty–sandy fillings in clayey–silty sediment, with mostly less than one whorl preserved, in which case the trochospiral shape may be inferred by change of level in the sediment (Figs. 34C, 35). Spiral best seen in rare specimens preserved in sandstone, filled with clayey material now often eroded and leaving hollows (Fig. 36C–D). The diameter of the spiral in most specimens tapers downwards. The maximum number of whorls observed is four, but no specimen is complete. Radius of spiral 2–11 mm. Cross-section of burrow typically oval or flattened, but in a few it is circular or near circular. Diameter of burrow 2–13 mm. Surface smooth or set with ridges about normal or slightly inclined to tangent of burrow. Spacing of ridges typically 0.5–1.0 mm. A few specimens have gradually pointed terminations. These do not show signs of being broken or eroded and may represent true lower endings. The spiral may occur in great numbers, and several burrows may be intertwined; a few show multiple burrows along part or whole of same spiral.

Material in burrow often rich in mica, concentrated to the outer part of the burrow (Fig. 35B, C). *Gyrolithes polonicus* occurs together with non-spiral burrows of similar diameter and general appearance, including some with mica-flake dense walls, although no direct connection have been found.

Taxonomy of vertical spiral burrows. – As *Spiroscolex* Torell, 1870, vertical spiral burrows have been recorded from the Mickwitzia sandstone since the 19th century but have remained poorly known. The first mention was in a report given by Otto Torell in Christiania (Copenhagen) in 1868 (*Forhandlingar ved de Skandinaviske Naturforskeres tiende Mode i Christiania 1868,* p. LXVII). The name, written as *Arenicolites spicalis,* is unaccompanied by any description and is a nomen nudum. The first published description of the species is by Linnarsson (1869a), and not, as previously given (e.g., Häntzschel 1975, p. W188), Torell (1870). Torell (1870) placed the fossil in a new genus, *Spiroscolex,* with the following diagnosis of *Spiroscolex spiralis* 'Spira duplex, diam. 20 mm; verme convoluto Trichinam spiralem mirum quantum imitante, annuli paullulum elevati, facile conspicui, conferti, 6–8 in spatio 10 mm., lat. maxima 5 mm'. Torell's (1870) description was not accompanied by illustrations, and subsequently the species has been rarely illustrated. Possibly only two slabs with *Spiroscolex* have been figured; originally by Linnarsson (1871) and Nathorst (1881a); later reproduced in, e.g., Walcott (1890, Pl. 61:6), and Häntzschel (1975, Pl. 108:3). The illustrated specimens are not very typical for the Mickwitzia sandstone spiral, and *Spiroscolex* has generally not been recognized as a vertical spiral, though this was stated by the early workers,

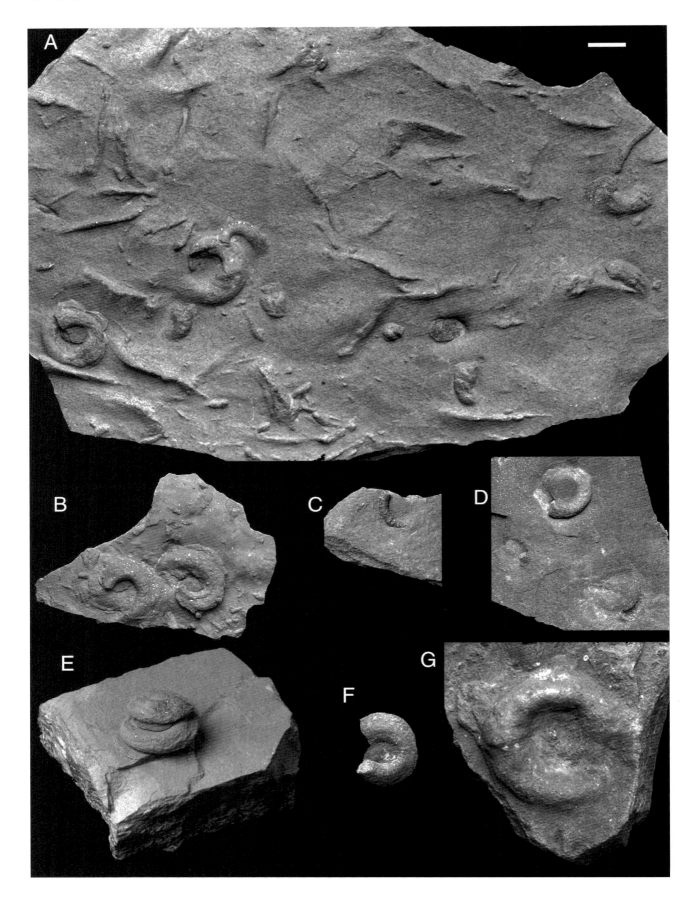

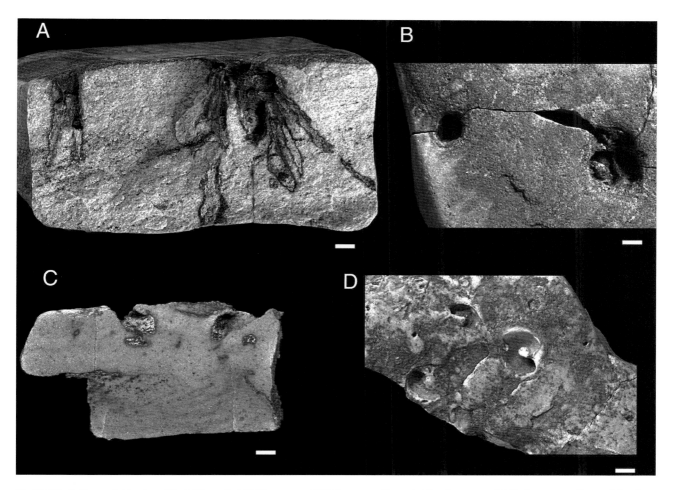

Fig. 36. □A, B. Top and side view of *Scotolithus mirabilis* Linnarsson, 1871. Burrow-walls marked by clayey material. Trolmen, top of interval A or base of interval B. RM X3346a, b. □C, D. *Gyrolithes polonicus* seen in top (C) and side (D) view. In the section the spiral is seen as elongated specks. Scale bar 10 mm. Trolmen, top of interval A, or base of interval B. RM X3306.

most elaborately by Linnarsson (1871). Häntzschel (1975) placed *Spiroscolex* in 'Unrecognized and Unrecognizable "Genera"'. As suggested by its name, *Spiroscolex* was originally thought to be a worm (Torell 1870; Linnarsson 1869a, 1871). Later Nathorst (1881a, b) suggested the possibility that it was a burrow fill or that it originated from medusa-tentacles. Martinsson (1974, p. 245) mentioned it under burrow fillings.

Fig. 35 (opposite page). *Gyrolithes polonicus* from the Mickwitzia sandstone. Scale bar 10 mm. □A. Several specimens on base of slab with fluted tool marks. Lugnås. PMU Vg 990. □B. Two flattened specimens with transverse fine annulations. Lugnås. SGU 8587. □C. Specimen with mica-rich surface and distinct transverse ridges. Lugnås. SGU 8588. □D. Specimens with transverse annulations. SGU 8589. □E. Specimen exhibiting about two whorls of the spiral. SGU 8590. □F. Specimen with faintly developed transverse annulations along the inner margin. Lugnås. SGU 8591. □G. Poorly preserved specimen that may be a *Spiroscolex crassa* Torell, 1870. Lugnås. SGU 8592

Torell (1870) also described a second species, *Spiroscolex crassus* Torell, 1870, which was characterized as being larger than *Spiroscolex spiralis*, having less pronounced annulations and only a single convolution. He gave its diameter of burrow as 13 mm and diameter of whorl as 35 mm. This species is problematic; the type material has not been found, and it has never been illustrated. While *Spiroscolex spiralis* clearly is a vertical spiral burrow, the description of *Spiroscolex crassus* does not necessarily indicate the same. Museum material is inconclusive; in the collections of the Swedish Geologic Survey, Uppsala, there are four specimens in a box labelled '*Spiroscolex crassus* Tor?'. Of these, one specimen cannot be confirmed as a trace fossil and another is in the shape of a horizontal U. The remaining two could be parts of trochospirals but may also be U-shaped (Fig. 35G). The label, written by von Schmalensée, states that according to a message from Professor Torell, true *Spiroscolex crassus* are somewhat thicker than these and have a smooth surface. A specimen of *Spiroscolex spiralis* collected from Billingen is similar to these two specimens but is a more likely case for a spiral. Its dimensions (diameter of burrow 9 mm, diameter of 'whorl' 32) attain those given by Torell for *Spiroscolex crassus*. Nathorst (1881b) said that *Spiroscolex crassus* occurs as fillings of spiral impressions on sediment partings, whereas *Spiroscolex spiralis* is found free. *Spiroscolex crassus* may constitute large specimens of *Spiroscolex spiralis*. However, the inhomogeneous nature of the available museum material and the unsuccessful search for type material renders *Spiroscolex crassus* a nomen dubium. Reports of *Spiroscolex* outside Sweden are scarce. It has been reported from the Lower Cambrian Lükati beds of Estonia (Öpik 1956) and Lower Ordovician Wabana Group of eastern Newfoundland (Seilacher 1970, p. 462; age according to Bergström 1976). The identity of a Precambrian form from Newfoundland reported as *Arenicolites spiralis* by Billings (1872) and compared to the Swedish specimens, can, according to Hofmann (1971, p. 18), no longer be ascertained. Bednarczyk & Przybyłowicz (1980) reported *Spiroscolex spiralis* from cores of the Lower Cambrian Leba Formation of northern Poland, but the identification is doubtful.

The most widely reported vertical spiral ichnogenus is *Gyrolithes* de Saporta 1884; as diagnosed by Bromley & Frey (1974, pp. 317–318) containing spiral burrows with a vertical axis. It is known mainly from marginal marine environments of mostly Mesozoic and (especially) Cenozoic age (e.g., Bromley & Frey 1974; Gernant 1972; Kilpper 1963; Mayoral 1986). There are several other ichnogenera of vertical spirals but these are probably better put in synonymy. *Conospiron* Vialov, 1969, from Oligocene beds of Tadzhikistan, has a conical course of its spiral (Vialov 1969). The conical profile of the spiral in *Conospiron babkovi* (Hecker, Osipova & Bel'skaya, 1962) is low (Hecker *et al.* 1962, Pl. 22:4; Vialov 1969) and is considered an accessory feature. *Megagyrolithes* Gaillard, 1980, is a large spiral of 30 cm diameter, consisting of a slender burrow, some centimetre thick, which is surrounded by a calcite capping, 20–25 cm in diameter. It was said to differ from *Gyrolithes* by its large size and mode of fossilization (Gaillard 1980) but is probably better regarded a species or preservational type of the same. Large vertical spiral burrows from continental deposits, generally found to be made by vertebrates, including therapsids (Smith 1987) and rodents (e.g., Lugn 1941) are known as *Daimonelix* Barbour, 1892. Generally they differ from *Gyrolithes* only by being larger and being found in continental deposits. Another spiral, *Ichnogyrus* Bown & Kraus, 1983, though considerably smaller than most *Daimonelix*, was also thought to be made by vertebrates, because of its occurrence in fluvial sediments. It was said (Bown & Kraus 1983) to differ from *Gyrolithes* in having a tightly coiled spiral with contacting whorls. The use of size and environment of production at the ichnogeneric level is not satisfactory, and all vertical spirals would best be included

in one ichnogenus, *Gyrolithes*. *Spiroscolex*, here considered a senior subjective synonym, has priority of age to *Gyrolithes* but for reasons of taxonomic stability (ICZN, Article 23b) it is proposed that *Spiroscolex* be suppressed.

Over fifteen ichnospecies of vertical spiral trace fossils have been described. To identify meaningful taxa requires study of type material and is beyond the scope of this work. However, the Mickwitzia sandstone spirals appear to be identical to *Gyrolithes polonicus* Fedonkin, 1981, as described from its type area, the Lower Cambrian of Poland. The descriptions and illustrations of specimens from eastern Poland (Fedonkin 1977, 1981; Paczesna 1985, 1986) are of burrows virtually identical to the Mickwitzia Sandstone specimens in shape, size and occurrence in the sediment. Also, in the Polish material there are specimens with annulate walls (Paczesna 1986, Pl. 3:3; J. Paczesna, written communication 1989), though smooth walls seem to be more common. The vertical spiral burrows described above are therefore assigned to *Gyrolithes polonicus*.

Gyrolithes polonicus differs from other species of *Gyrolithes* by its small size, lack of terminal expansion, and lack of scratch marks and carapace impressions. A downward taper of the spiral is common but not exclusive. Characteristic, but of doubtful taxonomic significance, are the transverse ridges whose origin may be inorganic (see below). *Gyrolithes saxonicus* (Häntzschel, 1934) is similar, and Fillion & Pickerill (1990a) suggested that they may prove synonymous. However, *Gyrolithes saxonicus* seems to differ from *Gyrolithes polonicus* in greater size, higher ratio of spiral diameter to burrow diameter and downward expanding width of spiral (see Häntzschel 1934, 1935; Richardson 1975). Specimens assigned to *Gyrolithes saxonicus* from the Upper Cambrian? – Lower Ordovician Bell Island and Wabana Groups of Newfoundland (Fillion & Pickerill 1990a) are within the size range of *Gyrolithes polonicus* but have a downward increase in width of spiral.

Vertical spiral burrows were reported from the Upper Breivik Member, northern Norway (Banks 1970). These were said to include irregular forms as well as specimens with a constant whorl diameter and a tube diameter mostly less than 5 mm (Banks 1970, p. 30). *Gyrolithes* has subsequently been reported also from the Lower Breivik (Farmer *et al.* 1992), but no additional information is available. *Gyrolithes polonicus* also occurs in the Torneträsk Formation of northern Sweden (Jensen & Grant, unpublished). In the East European Platform, *Gyrolithes polonicus* is, beside Poland, also known from the Rovno horizon of Podolia, Ukraine (Fedonkin 1983, p. 134). Material in the Palaeontological Institute, Moscow, collected by Dr. M. Fedonkin, showed them to have a burrow diameter of 2–3 mm and a spiral radius of 8–9 mm. Transverse ridges were only observed on one specimen. Probably to be included in *Gyrolithes polonicus* is *Gyro-*

lithes sp., reported by Liñan (1984) from the Precambrian – Lower Cambrian Torrearboles Formation of Spain. Material of *Gyrolithes polonicus* from North America is difficult to evaluate. Crimes & Anderson (1985) reported two specimens from the Chapel Island Formation of southeastern Newfoundland. They are sand-filled positive hyporeliefs, forming almost complete circles (Crimes & Anderson 1985, p. 321, Figs. 6:7 and 8), and the spiral of those specimens was inferred rather than observed. Landing *et al.* (1988) also reported 'Gyrolithes?' and 'Gyrolithes' from the Chapel Island Formation, but no additional information on morphology was given. The published evidence for the assignment of these trace fossils to *Gyrolithes* is not entirely convincing, and the specimens presented by Crimes & Anderson (1985) may pose the same interpretational problem as *Spiroscolex crassus*.

Fritz (1980) listed *Gyrolithes polonicus* from map unit 12 in the west-central Mackenzie Mountains, Canada.

Interpretation. – The fill of *Gyrolithes polonicus* is suggestive of having been passively introduced into an empty burrow. The sediment of the burrow is distinctly different from the surrounding material and lacks such typical feeding-trace characters as back-filling. The concentration of mica flakes to the burrow wall is indicative of a stabilizing burrow lining. Though this could point to a pure dwelling structure, it may also have been a dwelling–feeding structure, possibly a connection between the sediment surface and a preferred level of feeding.

Cenozoic specimens of *Gyrolithes* have mostly been attributed to decapod crustaceans (Hecker *et al.* 1962; Kilpper 1962; Keij 1965; Kennedy 1967; Gernant 1972; Mayoral 1986). Indices for this include long markings running parallel to the burrow wall, thought to be imprints of the carapace, and transitions to the burrow systems *Ophiomorpha* Lundgren, 1891, and *Thalassinoides* Ehrenberg, 1944, generally thought to be made by callianassid crustaceans. *Gyrolithes polonicus*, as far as known, lacks these characters, and also their age makes such producers unlikely (Häntzschel 1975). Toots (1963) discussed the pattern of movement and behaviour leading to formation of helical burrows, and found bilateral symmetry and paired appendages to be necessary features of the producing organism. Besides arthropods, also annelids meet these criteria, and Powell (1977) argued for capitellid polychaetes as the most likely producers of *Gyrolithes* burrows. In fact, Powell (1977) deemed the evidence for crustacean affinity of uniformly spiralling traces as untenable, because the spirals of modern crustaceans are much less regular, and intersections with crustacean burrows were considered as mere chance associations. Frey & Wheatcroft (1989, p. 269) mentioned spiraled burrows of certain capitellid polychaetes under the heading monotypical styles of bioturbation. Though this should not automatically be applied to the Cambrian spirals, a polychaete is a likely producer. Hertweck (1968, Fig. 13,

Pl. 2:2) reported spiral escape traces formed by the gastropod *Hydrobia ulvae* embedded by storms. In case of successful escape, the traces filled with sand. Those spirals appear to consist of closer spirals.

The transverse burrow-wall annulations could have resulted from burrowing movements of the producer. According to Linnarsson (1871), the absence of these ridges in some specimens may be due to adherence of fine-grained sediment to the sand-filled tubes. However, there is a tendency for ridges to be most prominent in specimens with oval cross-section, pointing toward a diagenetic origin, or at least enhancement, of the ridges. This could have resulted from different relative movement of the muddy and sandy–silty material during compaction, which would act strongly upon a full-relief endogenic burrow. As the ridges are less well developed on specimens with roundish cross-section, an inorganic origin must be considered. Their formation may then be similar to that of corrugations in 'Rhysonetron'-type mud-crack fillings (Young 1967, Fig. 4, 1969; Hofmann 1971). Similar ridges are also seen on *Gyrolithes* from the Devonian of Spain (Garcia-Ramos 1976, Pl. 1c) and on an irregular Cretaceous spiral trace from Germany (Häntzschel 1934, Fig. 4). The origin of all these ridges may be inorganic. The ethological motivation for digging a spiral burrow is not altogether clear. It could be seen as improved space utilization by a mud-eater. However, in the burrow system of the capitellid polychaete *Notomastus latericeus*, spiral segments of various orientation form only a part of the burrow system, and its function was given as unknown by Reineck *et al.* (1967, p. 243). Possible explanations include: asymmetry of the animal's digging organs; incapability of the animal to dig and move in a vertical stance; less deep burrow needed to conceal an elongated organism, thus possibly avoiding hardened sediment and adverse environmental conditions. In view of the likely association of the spirals to straight segments, the first possibility is less likely. Britton & Morton (1989, p. 263) noted that when the polychaete *Glycera americana* was placed in water it performed a spiral-shaped swimming motion which continued also after contact with sediment, and evidently functioned to screw the animal into the sediment.

Ichnogenus *Helminthoidichnites* Fitch, 1850

Helminthoidichnites tenuis Fitch, 1850

Figs. 27B, 37C

Material. – Eight slabs with hundreds of specimens; figured slabs RM X3298–99, 3308. Found in Interval B at Lugnås and Kinnekulle.

Description. – Straight to gently and irregularly winding burrows preserved as positive or negative hyporeliefs (Figs. 27B, 37C). Material in burrows structureless, similar to or coarser than surrounding sediments. Surface of burrow smooth or grainy, reflecting coarse material, rarely somewhat annulate. Width ranges from 0.2 to 2 mm, typically with dominance of a restricted size range on a given surface. Burrows in the lower size range may occur in large numbers with frequent crossing but only rarely with the same burrow crossing its previous path (Fig. 27B). Larger burrows typically have two shallow depressions bordering the burrow.

Discussion. – Häntzschel (1975) placed *Helminthoidichnites tenuis* Fitch, 1850, in *Gordia* Emmons, 1844, but it was maintained as a separate genus by Hofmann & Patel (1989), as it typically does not possess crossings of the same burrow, now considered diagnostic of *Gordia*. According to Hofmann (1990), *Helminthoidichnites* is characterized by a random movement-pattern whereas the type species, *Gordia marina*, has segments of nonrandom behaviour. He suggested, however (Hofmann 1990, pp. 18–19), that the two may be end members in a continuous spectrum. *Helminthoidichnites* differs from *Helminthopsis* Heer, 1877, in possessing less winding and less systematic meandering (Hofmann & Patel 1989), and Narbonne & Aitken (1990) also considered the presence of level crossings as distinguishing from *Helminthopsis*. The distinction from *Planolites* is, theoretically, the absence of lithological difference from parent rock as the result of deposit-feeding activities. However, this difference is very difficult to recognize in burrows of submillimetric size. *Helminthoidichnites* is common in upper Proterozoic strata and ranges throughout the Phanerozoic (Hofmann & Patel 1989; Narbonne & Aitken 1990). *Helminthopsis filiformis* Alexandrescu & Brustur, 1987, from Cretaceous–Paleogene of the Carpathian flysch (Alexandrescu & Brustur 1987), appears identical to *Helminthoidichnites tenuis*.

The overall similarity in course within the size range is taken to merit inclusion in the same ichnospecies. Most burrows are intergenic, made along a sand–mud boundary. Bordering grooves along the side of the burrow is a common trait of postdepositional traces (Osgood 1970). The purpose of the burrows may have been feeding or mere propulsion (Narbonne & Aitken 1990).

Ichnogenus *Hormosiroidea* Schaffer, 1928

'Hormosiroidea' isp.

Figs. 37A–B, ?34A

Material. – Six slabs; figured RM X3307, 3226; PMU Vg 989. Found in interval B at Kinnekulle and Lugnås.

Description. – The best preserved specimen consists of an about 1 mm wide, horizontally gently curved burrow, along one side bordered by vertical knobs about 1 mm wide (Fig. 37B). Other knobs are largely covered by the horizontal burrow. Spacing between the centers of vertical knobs is about 4 mm. Other specimens consist of vertical knobs arranged in straight to curved lines with knobs separated or cutting into each other (Fig. 37A). Lower surface of knobs flat, horizontal or inclined. Occasionally the knobs turn into an undulating horizontal burrow (Fig. 37A). Width of knobs about constant within a series; in two specimens 0.6 and 2 mm, respectively. Still other burrows consist of gently to acutely convex, elongate segments arranged in a straight to curving series where each segment generally is separated from the one preceding and following. Width of segments 0.5 mm, horizontal length of segments 1–4 mm.

All specimens are found on bases of lenticular silt–sandstones from interval B at Kinnekulle.

Discussion. – Schaffer (1928) erected *Hormosiroidea* for spherical or hemispherical bodies connected by a narrow string-like structure. There are several reports of Cambrian *Hormosiroidea*, but these appear to be different from Schaffers material. Uchman (1995) assigned to *Saerichnites* forms typically placed in *Hormosiroidea*, but it is not clear that *Saerichnites abruptus* Billings, 1866, is a burrow of the type described here. The reconstruction of *'Hormosiroidea'* is problematic as it reveals only a plane through or half of the structure (Seilacher 1977; Crimes & Anderson 1985). Crimes & Anderson (1985) gave two alternative interpretations for Lower Cambrian specimens from Newfoundland; one with a vertically meandering burrow and one with a main horizontal burrow with vertical shafts. The latter model seems to apply to the most well-preserved specimen and could also serve for the rest. The burrow was probably formed by a main horizontal, gradually progressing burrow from which a vertical shaft was made at intervals. The purpose of the vertical shaft was probably to retain contact with well-oxygenated water and possibly also for defecation. If this interpretation for *'Hormosiroidea'* and the analogous interpretation for *Treptichnus* are correct, these are basically similar building units, though added at different angles (Fig. 23).

Fig. 37. □A. '*Hormosiroidea*' isp. In top part of picture are several small knobs arranged in rows, occasionally with partial truncations. In lower left corner is a larger specimen. Scale bar 5 mm. Hjälmsäter. RM X3307 □B. '*Hormosiroidea*' isp. Specimen preserving both vertical knobs and horizontal burrow positioned below the knobs. Scale bar 5 mm. Hjälmsäter. Interval B. RM X3226. □C. Lower surface with bilobate *Cruziana tenella* and small simple *Helminthoidichnites tenuis*. Scale bar 10 mm. Hjälmsäter, Interval B. RM X3308.

Forms similar to the specimens described above, except for a smaller size, were reported by Bryant & Pickerill (1990) from the early Cambrian Buen Formation of North Greenland. Lower Cambrian specimens reported by Crimes & Anderson (1985) from Newfoundland, by Walter *et al.* (1989) from central Australia and by Bryant & Pickerill (1990) from north Greenland all have the burrows arranged in looping chains. Only short segments are preserved in the above described specimens, but they do indicate greater irregularity.

Ichnogenus *Monocraterion* Torell, 1870

The name *Monocraterion* is used for funnels penetrated by a central vertical, more or less straight tube (Häntzschel 1975). It has generally been interpreted as a vertical dwelling structure of a suspension feeder. The relation between *Skolithos* (consisting of simple vertical tubes) and *Monocraterion*, and their possible synonymy have repeatedly been addressed. Westergård (1931a, p. 12) noted that incomplete specimens of *Monocraterion* are indistinguishable from *Skolithos*, and several authors have considered the two to be made by the same or similar animals under different environmental conditions. Hallam & Swett (1966) interpreted the funnels in *Monocraterion* to result from escape movements of the animal caused by sediment influx, comparing the burrow with that of modern sea anemone *Cerianthus*. Goodwin & Anderson (1974), studying material from the Cambrian Chickies Quartzite, considered the similarity superficial, instead pointing to evidence that the funnel was made by the tube dweller, possibly as a consequence of feeding motions. Studying distribution of burrows of the polychaete *Diopatra cuprea* on present-day point bars, Barwis (1985) found funnel-topped burrows restricted to the upper part of the bar, consisting of poorly sorted, thin-bedded sediments. In *Diopatra cuprea* the funnel forms as a scour pit (Barwis 1985). Durand (1984, 1985a) discussed the relation between *Monocraterion* and *Skolithos* in the Ordovician Gres Armoricain in France. Durand (1985a) observed that *Monocraterion* and *Skolithos* were typically found in different trace-fossil assemblages and environments. The name *Skolithos* was used for long tubes of constant diameter found in homolithic sandstone facies. The upper part of these burrows occasionally has a funnel-shaped top; its preservation depends on erosive or non-erosive sedimentation (Durand 1984). In *Monocraterion*, found with *Cruziana* in heterolithic facies, the funnel is more elongated, which was suggested to imply none or little sedimentation (Durand 1985a).

There could thus be several origins of funnel-shaped tops. Bromley (1990) considered *Monocraterion* a junior synonym of *Skolithos*, but several authors continue to consider the funnel-shaped tops as a characteristic and useful feature upon which to base ichnogenera (Alpert 1974b; Fillion & Pickerill 1990a, b). Fillion & Pickerill (1990b) illustrated a slab supposed to have both *Monocraterion* and *Skolithos*. Their specimen of *Monocraterion* is a gently flaring conical structure. This does not exclude the possibility that it is the upper part of a deeper structure, the more parallel-sided lower end of which would be called *Skolithos*. Furthermore, the morphology of their specimen grades into plug-shaped burrows such as *Conichnus*. A further problem is that *Monocraterion* could form around the vertical parts of an infaunal burrow system as well as around a simple vertical dwelling structure.

Several burrows with conical tops in the Mickwitzia sandstone can be seen to curve to a horizontal position within the sediment, and some form the top of a radiating burrow system. The problem is related to bedding thickness and the state of preservation and the amount of material preserved. The vertical parts of an extensive burrow system such as *Thallasinoides* are not named separately (but see discussion on *Rosselia*). When a vertical burrow with a funnel-shaped top is found with a broken base, it cannot be decided whether it was a simple vertical structure or part of an extensive burrow system, or even a plug-shaped burrow. This uncertainty applies also to the type material of *Monocraterion tentaculatum* described below. Frey & Howard (1985) stated the differences between *Monocraterion* and *Rosselia* to be that the basal stem is simple in *Monocraterion* but concentrically laminated in *Rosselia*; whereas stacked funnels in *Skolithos* reflect physical processes, that of *Rosselia* is said to reflect sediment processing by a deposit-feeding animal. This interpretation of *Rosselia* must be questioned, as in several occurrences (the present work; Rindsberg & Gastaldo 1990) and probably also in the type area for *Rosselia socialis*, it more probably represents reactions to sediment influx (see below). A funneled top is morphologically distinctive and its presence or absence gives information on sediment consistency and depositional environment, but the uncertainties regarding its formation make it a doubtful criterion on which to base ichnotaxa.

Whether *Monocraterion* and *Skolithos* should be synonymized is thus a question of having one or two purely descriptive headings, which in either case include forms of different original morphology.

Monocraterion tentaculatum Torell, 1870

Figs. 38, 39A, 40A

Synonymy. – □1870 *Monocraterion tentaculatum* n.g. et sp. – Torell, p. 13. □?1891 *Monocraterion magnificum*, n.sp. – Matthew, p. 161, Pl. 26.

Material. – Two slabs collected and figured, SGU Type 5364, RM X3309.

Description of type material. – A slab with number SGU Type 5364 in the collection of type material at the Swedish Geological Survey is most probably among the material studied by Torell (Figs. 38, 39A). The slab consists of fine- to medium-grained sandstone, is about 60 mm thick, and has a flat base and a rippled (interference type) top. Four funnel-shaped trace fossils are seen on the upper surface, of which two are visible in vertical section at the slab margin. At the floor of the funnel are raised knobs. These are continuous with vertical rods of a smaller diameter, going down into the sandstone. In one of the specimens, numerous small burrows go out from the central knob

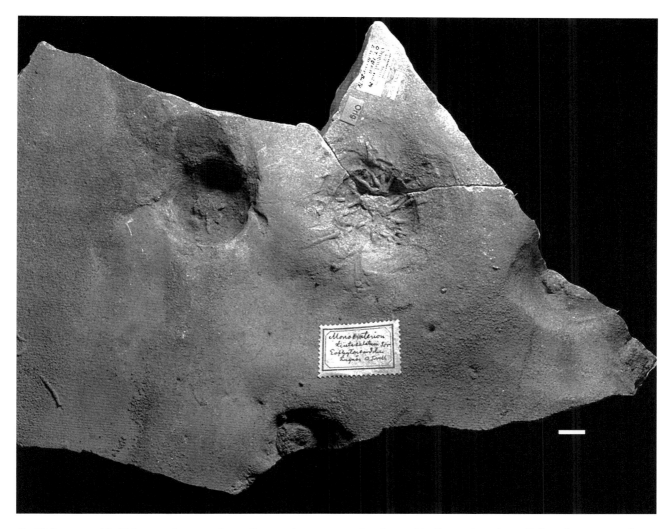

Fig. 38. Type material of *Monocraterion tentaculatum*. Four specimens on top of sandstone slab. One specimen has radiating tubes corresponding to Torell's (1870) description of tentacles. Scale bar 10 mm. Lugnås, probably interval A. SGU Type, 5364. See also Fig. 42.

(Fig. 39A). These have a diameter of 1.3–1.6 mm, consist of finer-grained sandy and clayey–silty material, and stand out by being grayish as opposed to the reddish-brown sandstone. The small burrows have a smooth outer surface, and some specimens show a distinct wall lining. This specimen is here designated lectotype

Discussion. – The whereabouts of the material upon which Torell (1870) described *Monocraterion tentaculatum* was previously not known. Westergård (1931a) considered specimens from the lower part of the Lingulid sandstone (here considered the upper part of the Mickwitzia sandstone) as possibly similar, and these have since stood as models for *Monocraterion*. A major remaining inconsistency has been that in Torell's (1870) description, *Monocraterion tentaculatum* is said to possess tentacles, numbering about 20, with lengths up to 34 mm. Nathorst

(1881b, p. 50) thought that the tentacles were cemented threads of excrement. Westergård (1931a) failed to find structures resembling these tentacles in his material. Also, the measurements of the funnels were at variance; Torell's (1870) specimens had a width of 30–40 mm, whereas the specimens described by Westergård (1931) have a width of 10–15 mm, rarely 20 mm.

The specimen described above fits Torell's (1870) description well. According to the label it originates from the Mickwitzia sandstone, and from the character of the rock it can be assigned to one of the basal beds. Funnel-shaped burrows are common in several of the basal beds at Trolmen. These are generally found in a weathered state, which may account for the absence of the radiating burrows. Funnel-shaped burrows surrounded by radiating impressions are also found in interval C. This includes a specimen on slab RM X3309 (Fig. 40A), 33 mm wide at

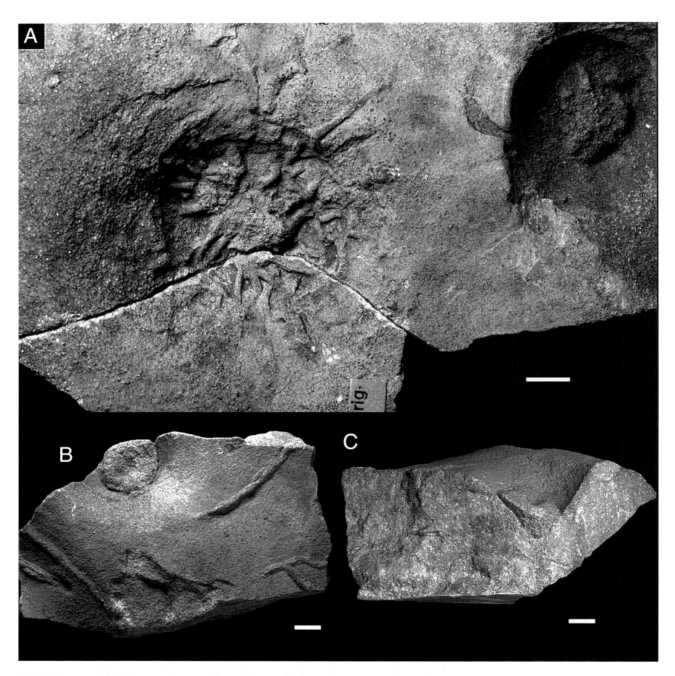

Fig. 39. Type material of *Monocraterion tentaculatum.* □A. Detail of specimen in Fig. 41, showing burrows radiating from an elevated central knob. Scale bar 5 mm. □ B, C. *Monocraterion* isp. seen in upper (B) and side (C) view. According to museum label this was probably also among Torell's type material. In side-view a vertical burrow is seen below the wide, partly filled funnel. Scale bars 10 mm. SGU Type 5363.

the top and of about the same depth, which is surrounded on the only preserved side by straight to gently curved, about 40 mm long burrows, preserved in negative epireliefs. There are about eight ridges covering one quadrant around the funnel.

Interpretation. – From the presence of a lining, the radiating burrows appear to be permanent or semiperma-

nent structures rather than impressions left by foraging tentacles. They may represent excursions of the animal living in the vertical tube (Fig. 21C). The diameter of these tubes is small compared to the vertical funnel-shaped structure, though this is to be expected. Many organisms in dwelling burrows will under conditions of gentle sediment input adjust to this by pressing sediment to the walls, and the burrow size will be considerably

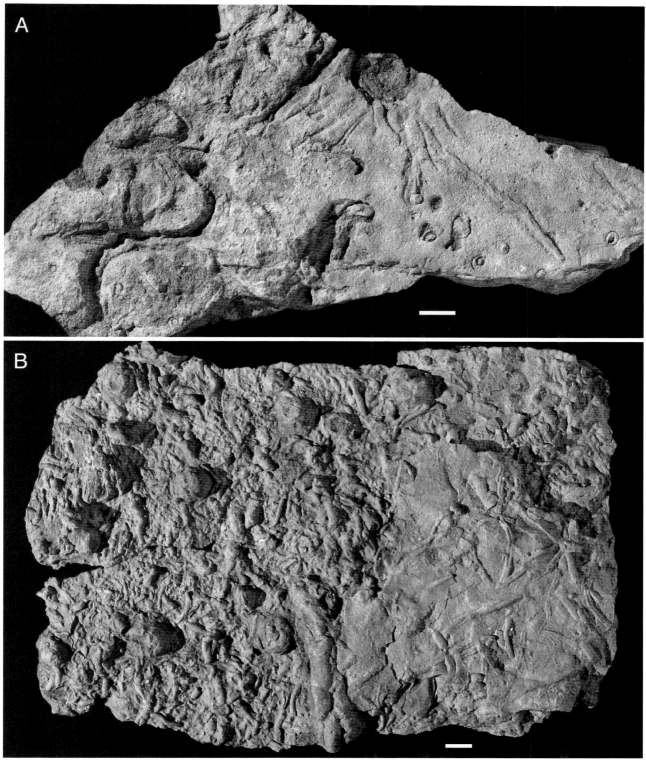

Fig. 40. □A. *Monocraterion tentaculatum* with poorly developed radiating ridges. In lower part of picture are also seen filled vertical tubes, probably *Monocraterion*, and small fragments of an inclined spreite burrow, probably *Rhizocorallium jenense*. The horizontal part of a *Rhizocorallium jenense* spreite is seen on the left side of the picture. Scale bar 20 mm. Älerud, Lugnås. From the upper part of the Mickwitzia sandstone. Similar specimens observed in beds 6 m above basement. RM X3309 □B. *Monocraterion* cf. *tentaculatum.* Several specimens preserved as round knobs with a central cylindrical structure, probably the fracture surface of a vertical tube. Also seen are abundant small *Palaeophycus tubularis.* In lower part of picture is a *Phycodes* cf. *curvipalmatum.* The base of an underlying bed (lower right side), has numerous *Cruziana tenella.* Scale bar 10 mm. Älerud, Lugnås. 2.7 m above basement. RM X3310.

larger than the animal itself. Material filling the central living tube typically has weathered less easily, as it consist of cleaner sand. In some funnel-shaped burrows near the base of the Mickwitzia sandstone it is common for the central tube to be selectively preserved and rest at an angle over the funnel, possibly what Torell (1870) meant by 'when the worm reaches the upper part of the funnel, it seems to thrive in a sheltering tube, as is the case of recent worms of the division Tubicola' (translation from Westergård 1931a, p. 11).

Monocraterion cf. *tentaculatum* Torell, 1870

Figs. 40B, 41

Synonymy. – □(?*pars*) 1870 *Micrapium erectum* n.g. et sp. – Torell, p. 11.

Material. – Figured slabs, RM X3310–12; PMU Vg 991. Additional specimens in the collection of SGU. Found in thin-bedded sediment of heterolithic character at Lugnås and Hällekis.

Description. – Elongated conical to pyramidal, vertically oriented burrows, with or without an upper, flat to somewhat conical, symmetric, or often asymmetric collar (Figs. 40B, 41). Sides of collar smooth or with crude concentric pattern. Sides of conical burrow with fine longitudinal ridges, or with coarser ribs of less regular orientation which may continue onto the collar. In sections, cone is seen to penetrate collar (Fig. 41A). Base of burrow may have a pore-like invagination (Fig. 41E) or be extended into a nipple-shaped structure, though commonly the lower termination is not seen, owing to incomplete preservation. The trace is found as protuberances at the base of sandstone lenses less than a few centimetres thick or as free specimens within clayey beds (Figs. 40B, 41). The latter have a concave funnel-shaped top (Fig. 41C). Diameter is up to 10 mm, including collar up to 25 mm. Length up to 10 mm.

Discussion. – *Micrapium erectum* was erected by Torell (1870) for fossils from the Lower Cambrian of Västergötland, but it has never been figured and has only been briefly discussed (Nathorst 1881b; Matthew 1901; Westergård 1931a; Bruun-Petersen 1971). The type material has not been found, but specimens in the collections of the Swedish Geological Survey labelled *Micrapium erectum* agree with Torell's (1870) description. These burrows are elongated conical structures with smooth sides or fine longitudinal ridges and with an upper part having a slightly widened collar. They correspond in size and general morphology with Torell's description (1870, pp. 11–12). Nathorst (1881b, p. 50) doubted the biogenic origin of *Micrapium* and suggested it to be a gas-escape struc-

ture. However, Westergård (1931a) stressed similarity of *Micrapium* with *Monocraterion*. Though a definitive statement cannot be made without the type material of *Micrapium erectum*, it seems likely that burrows described above as *Monocraterion* cf. *tentaculatum* correspond to Torell's taxon.

These burrows could be interpreted in three ways: (1) as the upper part of a *Skolithos* or *Monocraterion* burrow, (2) as the vertical outlet of a more extensive burrow system, or (3) as the burrow made by a sea anemone-like creature whereby the elongated conical structure was formed by the penetrative part of the body whereas the collar was formed by tentacles in a later stage.

The most likely interpretation is that it represents the funnel-shaped top of a vertical burrow, formed in more thin-bedded sediments where the lower parts of the burrow usually are not preserved. Torell (1870, p. 12) mentions the fragment of a tube 60 mm long in connection with *Micrapium*. This appears to correspond to occasional specimens of *Palaeophycus* that have expanded upper vertical parts with a collar. The impressions on the funnel side probably represent outlets or impressions of burrows such as those seen on the type specimens of *Monocraterion tentaculatum*. Sections reveal that the flat collar is penetrated by a conical part (Fig. 41A). This outer collar most probably was shaped by scour; the same interpretation was also given for similar burrows from the Carboniferous of England (Eagar *et al.* 1985, Pl. 4E). The scour origin is also suggested by specimens having an asymmetric collar, with a steeper side closest to the vertical plug (Fig. 41C). There is a great similarity between the upper part of *Monocraterion* preserved in sandstone beds (Fig. 39B, C) and specimens formed in heterolithic sediments (Fig. 41C).

The possibility must also be considered that *Micrapium* is a plug-shaped burrow. Among plug-shaped burrows, general similarity exists with *Conostichus* Lesquereux, 1876, in which are typically included conical burrows with transverse constrictions and longitudinal ornamentation, and typically with an apical disk (Pemberton *et al.* 1988). The recently most favored interpretation of *Conostichus* is as an actinian dwelling burrow (Chamberlain 1971; Pemberton *et al.* 1988). In *Micrapium* the collar is, when present, proportionately wider and flatter than typical in *Conostichus*. Typical *Conostichus* have features such as the impression of an apical disc or a duodecimal symmetry. However, specimens lacking these characters have also been assigned to *Conostichus*. For example, Hakes (1976) reported *Conostichus* from the Upper Pennsylvanian of Kansas, consisting of small conical to subconical forms that occasionally had crude longitudinal ribs running from the apical end. Some specimens had small rod-like structures visible on the lower and upper end (Hakes 1976, p. 24). The question whether they could be funnels penetrated by a burrow was left unresolved, as sections

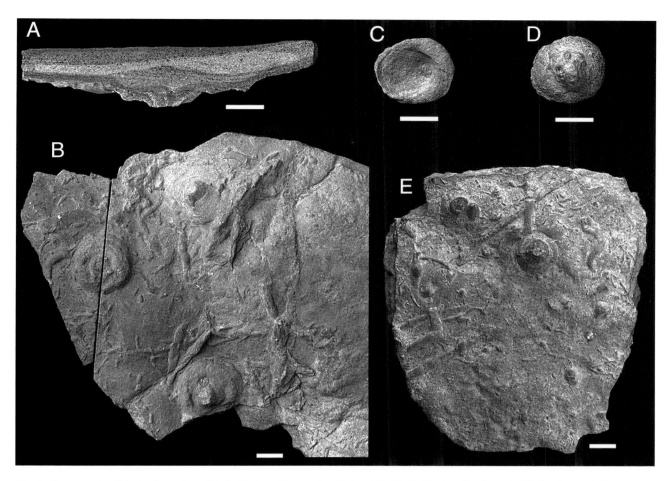

Fig. 41. Monocraterion cf. *tentaculatum* from thin-bedded sandstone units of interval D. □A. Polished section through slab showing vertical tube penetrating through a flat collar. The collar and tube are separated from the overlying sandstone bed by a clay seam. Scale bar 10 mm. Lugnås. PMU Vg 991. □B. Plane view of slab in A, with three specimens possessing a central conical part surrounded by a collar. Also seen are shrinkage cracks. Scale bar 10 mm. □ C, D. Loose specimen seen from above C, and below D. Note egg-shaped outline. Scale bar 10 mm. Hällekis, interval D. RM X3311. □E. One well-preserved and several indistinct *Monocraterion* cf. *tentaculatum*. At the center of the well-preserved specimen is an invagination surrounding a central broken plug. Scale bar 10 mm. Hällekis. RM X3312.

failed to reveal any such structure. It may prove very difficult to determine whether such specimens are plug-shaped burrows or assignable to *Monocraterion*. The distal ornamentation in a few specimens is similar to that of the holotype of *Bergaueria radiata* from the lower Cambrian of California (Alpert 1973, Fig. 3; Pemberton *et al.* 1988, Fig. 7A, B). One specimen preserved in full relief has a rounded lower surface with a sub-central depression from which coarse ribs radiate. The ribs are well developed only over half the surface. Diameter at top about 20 mm; length 10 mm. By its overall geometry, this burrow is similar to *Bergaueria*, though this ichnogenus does not have broken vertical structures.

Taxonomic implications. – The presence of radiating burrows (tentacles of Torell) was one of the main characters in Torell's (1870) description of *Monocraterion tentaculatum* and is a highly distinctive feature. This feature has a

low preservational potential, and its absence could depend on original absence or erosion. Still, it is so distinct that it merits retention as a species character. Comparison can be made with the large form *Monocraterion magnificum* Matthew, 1891, from the St. John Group of Canada (Matthew 1891), which also has tentacle-like structures radiating from a conical central structure. In *Monocraterion magnificum*, radiating imprints are found on the upper, flaring part of the burrow (Matthew 1891, 1901). Most assignments of *Monocraterion* have been to the ichnogeneric level only, a practice that will probably have to remain. Features such as degree of conicality will reflect the nature of the sediment and is therefore an inappropriate trait to base a taxon on. *Monocraterion rajnathi* Badve & Ghare, 1978, was characterized as having 'a very thin wall and wider central tube' (Badve & Ghare 1978, p. 128) in comparison to the Swedish material (of Westergård); these are doubtful characters for a new ichnospecies, though a final decision must await further description. *Monocraterion clintonense* (James, 1892) is in plane view similar to *Monocraterion rajnathi* and to poorly preserved *Monocraterion tentaculatum* (see Howell 1946, Pl. 2:1). *Monocraterion tentaculatum* can be compared with *Micatuba verso* Chamberlain, 1971, from the Carboniferous Atoka Formation of Oklahoma. *Micatuba* is a rarely reported form consisting of sand-lined tubes of 1–1.5 mm diameter, diverging radially on the top of rippled sandstone, interpreted to represent repeated probes from a central dwelling burrow (Chamberlain 1971, p. 238).

Here *Monocraterion tentaculatum* is restricted to specimens preserving radiating burrows. Specimens lacking such radiating burrows but presenting funnels are referred to as *Monocraterion* isp. (Fig. 39B–C). It must be stressed that this is as a purely descriptive scheme.

Ichnogenus *Monomorphichnus* Crimes, 1970

Monomorphichnus consists of sets of elongated narrow ridges, isolated or laterally repeated, generally thought to be made by sideways swimming–raking arthropods. Seilacher (1985, 1990) considers *Monomorphichnus* to be synonymous with *Dimorphichnus* Seilacher, 1955, which beside the raking imprints also have shorter, blunter imprints, called pusher marks. Though most reports of *Dimorphichnus* give the pushers as more deeply impressed than the rakers, Goldring & Seilacher (1971, p. 429, Fig. 2C) stated that the pushers in most *Dimorphichnus* are missing because of undertrack fallout, since the claws were spread out, acting as snowshoes. Seilacher (1955a, 1985) considered *Dimorphichnus* to be made by a trilobite stirring up and filtering sediment, with the hood of the animal kept close to the sediment to form a filter chamber. Other interpretations of this type of trace fossil have been given: Osgood (1970) suggested a trilobite caught in oscillatory currents. Current-swept trilobites, dead or living, were also considered by Martinsson (1965) and Banks (1970). Crimes (1970) interpreted *Monomorphichnus* as a trace made by a trilobite swimming immediately above the sediment surface, possibly affected by currents, raking the sediment with only one side of legs, possibly with the intention to rake up or disturb benthic animals in search for food. According to Seilacher (1985, p. 236), the holotype of *Monomorphichnus* (in Crimes 1970) have both pushers and rakers; according to Fillion & Pickerill (1990a, p. 40) it does not. As seen from the illustration of the holotype (Crimes 1970, Pl. 12C), each of the laterally repeated sets partly overlaps, with another set inclined at an angle. If these are genetically related the trace is of *Dimorphichnus*-type. The possibility that *Monomorphichnus* represents a behavior differing from *Dimorphichnus*

warrants continued separation. *Monomorphichnus* should not be confused with single-sided preservation of *Cruziana* and *Rusophycus*. It should be born in mind that similar structures can form inorganically as tool marks (see e.g., Allen 1982; Dżułyński & Sanders 1962), this the more as *Monomorphichnus* is commonly found on current-marked beds.

At least seven ichnospecies of *Monomorphichnus* have been erected, based on the number and grouping of ridges (e.g., Crimes 1970; Crimes *et al.* 1977; Alpert 1976b, Legg 1985; Fillion & Pickerill 1990a). Some species are defined on whether the ridges are paired or single. In the specimens from the Mickwitzia sandstone the morphology is dependant on the depth of the trace, the number of ridges increasing with depth. I suspect that a similar relation may exist also in some other occurrences, and therefore no species designation is made.

Monomorphichnus is commonly reported from Palaeozoic strata; e.g., Cambrian (Alpert 1976b; Crimes & Anderson 1985; Fritz & Crimes 1985), Ordovician (Aceñolaza & Manca 1982; Baldwin 1977), Silurian (Narbonne 1984). Most occurrences have been attributed to trilobites, but other arthropods could be considered as well; Romano & Melendez (1985) assigned Carboniferous specimens from Spain to eurypterids or xiphosurids. Jenkins *et al.* (1983) compared fan-shaped supposed arthropod scratch marks from the Vendian of Australia with *Monomorphichnus*. However, specimens illustrated by Gehling (1991, Pl. 6:3) that meet this description show that this form is different, including the fan-shaped arrangement of the scratches.

Momomorphichnus isp. A
Fig. 42

Material. – Five slabs; figured slabs RM X3313–14. Found at Lugnås and Kinnekulle, interval B.

Description. – Long shallow ridges, single or double, occurring in laterally repetitive sets (Fig. 42). Best preserved specimen (RM X3314) consists of about 13 elongated markings arranged in a parallel fashion but laterally displaced from each other (Fig. 42B). Distance between markings 2–10 mm. Most markings have a single ridge, but double ridges with a spacing of about 1 mm also occur. The markings are straight or slightly sigmoidal, with a length up to 50 mm. Within boundary of slab, the set is laterally repeated at least four times.

Discussion. – Seilacher (1955a, p. 91) mentions a slab with sigmoidal raking marks from the Mickwitzia sandstone in Lugnås, and though there were no pusher imprints, their belonging to *Dimorphichnus* was considered beyond doubt. The specimen (PMU number ar 2652) has not been possible to locate, but it appears to be similar to the

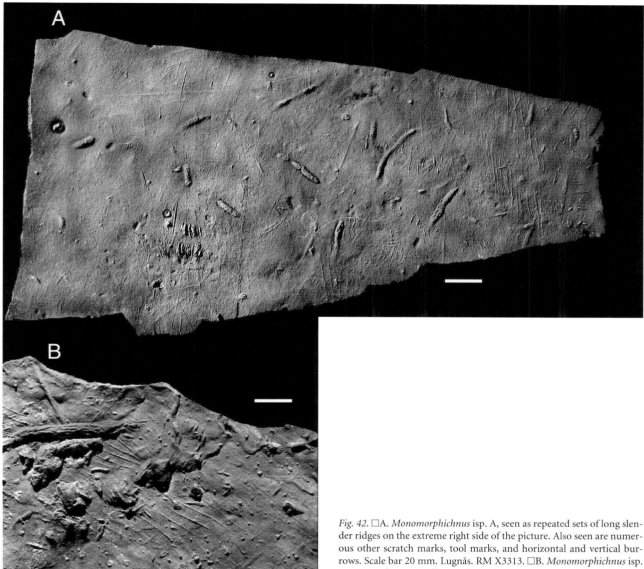

Fig. 42. □A. *Monomorphichnus* isp. A, seen as repeated sets of long slender ridges on the extreme right side of the picture. Also seen are numerous other scratch marks, tool marks, and horizontal and vertical burrows. Scale bar 20 mm. Lugnås. RM X3313. □B. *Monomorphichnus* isp. A, seen as three sets of long ridges. Also seen is *Palaeophycus imbricatus*. Scale bar 20 mm. Lugnås. RM X3314.

specimens described here. The absence of pushers is considered of importance, with the reservation given above, and they are assigned to *Monomorphichnus*. Interpretation of this form is hampered by the size of the slabs.

Monomorphichnus isp. B

Fig. 43, 67C?

Material. – Figured slab, SGU 8593. Lugnås, interval B.

Description. – Deep, straight, elongated markings, several arranged parallel to each other but each gradually shifted in longitudinal position from previous marking (Fig. 43). Each marking bears at least four longitudinally directed parallel ridges. On several markings there is a terminal shallow ridge, continuous with the main sets but set at a sharp angle. The most well-preserved set (slab SGU 8593) consists of about seven markings, 30–40 mm long, up to 3 mm wide and 2 mm deep. Another set on the same slab may be a repetition of the first set.

Remark. – *Monomorphichnus* isp. B differs from *Monomorphichnus* isp. A, in having shorter, deeper scratches. These deep rakings demonstrate the influence of depth on the number of ridges within a mark. This trace was deeply incised in the sediment; during compaction the deepest parts have in several cases folded over shallower parts. The claw pattern is identical to that on *Rusophycus dispar* and Arthropod trace Type A (see below).

Fig. 43. Monomorphichnus isp. B. On the right half of the picture are widely paired, short marks probably representing undertrack preservation of *Cruziana rusoformis.* Scale bar 10 mm. Lugnås. Probably interval B. SGU 8593.

Ichnogenus *Olenichnus* Fedonkin, 1985

Olenichnus isp.

Figs. 44

Synonymy. – □?1870 *Psammichnites filiformis* n.sp. – Torell, p. 10.

Material. – Seven slabs collected; figured RM X3315, SGU 8594. Found at Hjälmsäter, Lugnås and Billingen.

Description. – Vertical, inclined and most commonly horizontal burrows with a width of 1–2 mm. Horizontal burrows with irregularly winding, crudely meandering, or straight course, often forming irregular and incomplete networks (Fig. 44). A few burrows horizontally continuous for about 10 cm, but most are shorter. Junction between burrows often T-shaped, sometimes in the form of short appendages running from main burrow (Fig. 44A). Reburrowed segments occur. Preserved as negative hyporeliefs, often caused by subsequent removal of sediment below tunnel or by deflection of sediment into the tunnel. Occasionally formed as low intergenic ridges, which may have transverse constrictions, giving an annulate appearance (Fig. 44A). Longitudinal sections of segments with annulate walls have not revealed internal structures. In some specimens there are faintly developed vertical to inclined concentrations of clayey material possibly corresponding to a wall lining. Along short sequences the burrows are filled with a more coarse-grained material than surrounding sediment.

The vertical components are mostly seen as broken knobs at top and bottom of slabs; only rarely a direct continuation between horizontal and vertical burrows is visible.

Discussion. – This trace apparently consisted of vertical and inclined tunnels leading down to a system of horizontal burrows (Fig. 21A). The branching of the horizontal tunnels is sometimes seen to be false or secondary successive, while others, especially T-shaped junctions, probably are true simultaneous branchings (cf. D'Allesandro &

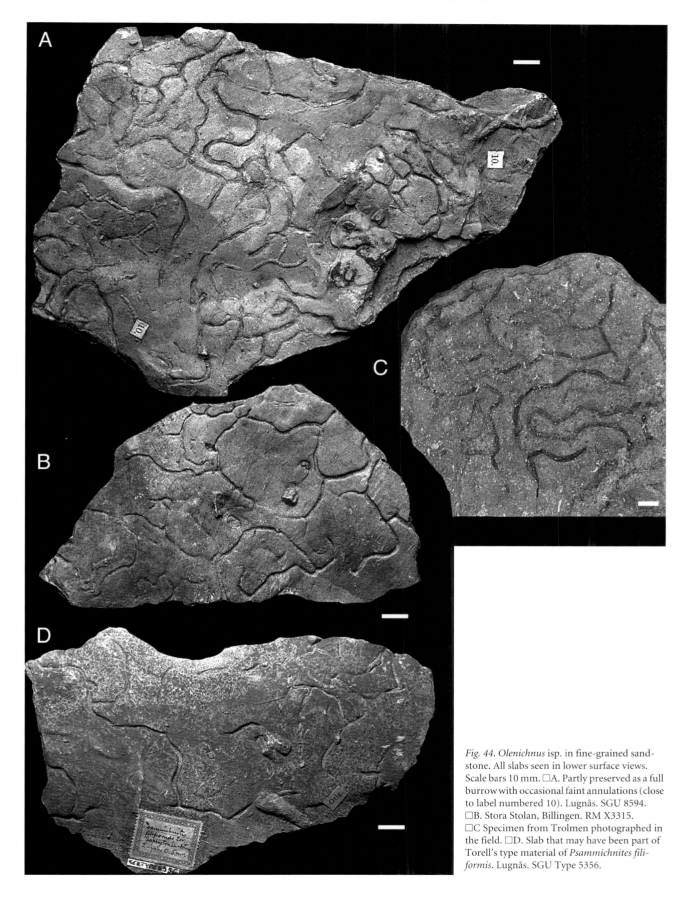

Fig. 44. Olenichnus isp. in fine-grained sandstone. All slabs seen in lower surface views. Scale bars 10 mm. □A. Partly preserved as a full burrow with occasional faint annulations (close to label numbered 10). Lugnås. SGU 8594. □B. Stora Stolan, Billingen. RM X3315. □C Specimen from Trolmen photographed in the field. □D. Slab that may have been part of Torell's type material of *Psammichnites filiformis.* Lugnås. SGU Type 5356.

Bromley 1987). The secondary successive branching may be due to reburrowing of partly collapsed or passively filled segments. The preservation of the trace suggests that it consisted of at least semi-permanent tunnels. Intrusion of sediment from below indicates some stability of the burrow, possibly by mucus binding. Annulate segments may reflect propulsion. These burrows were probably feeding–dwelling structures comparable to those made by polychaetes (see Howard & Frey 1975).

Taxonomy. – Torell (1870) erected *Psammichnites* to include a wide range of horizontal vermiform structures of which *Psammichnites gigas* was later chosen as type species by Fischer & Paulus (1969). In the type collection of the Swedish Geological Survey, there is a specimen numbered Type 5356, which the label says was found in Torell's work room and probably is a type specimen of *Psammichnites filiformis*. In the description of *P. filiformis*, Torell (1870, p. 10) mentions a width of 1.5 mm, eight annulations over a distance of 3 mm, and a meandering habit. The probable type specimen satisfies these general characteristics, though no annulations as dense as those reported by Torell (1870) are seen. Nathorst (1881b, p. 52) considered *P. filiformis* as probably the same as *P. impressus* and identical to *Fraena tenella*. As *Fraena tenella* has a marked median furrow this cast some doubt on the morphology of *Psammichnites filiformis*, and its relation to the traces described here is not certain.

Similar traces from the Nemakit–Daldyn Horizon of the Siberian Platform were described as *Olenichnus irregularis* by Fedonkin (1985, p. 206). The description gives no notion of vertical segments, but on the illustrated specimens (Fedonkin 1985, Pl. 23:2 and 24:2) there are several knob-like protrusions, possibly from vertical shafts. Such vertical shafts were also seen on specimens observed in the Palaeontological Institute, Moscow. The similarity in morphology is such that the Mickwitzia sandstone specimens are assigned to *Olenichnus*.

Crimes (1994, p. 107) claims that Fedonkin (1985) introduced *Olenichnus irregularis* and other new ichnotaxa from the Kessyuse Formation without giving a formal description. Hofmann & Patel (1989) considered that *Didymaulichnus meanderiformis* Fedonkin, 1985, described in the same paper as *Olenichnus irregularis,* was not properly diagnosed because it has no 'accompanying description or definition that would allow it to be differentiated from other taxa'. However, this is only a recommendation of the ICZN (Recommendation 13A) and not mandatory. There is accompanying text characterizing these taxa, and I see no reason to consider *Olenichnus irregularis* Fedonkin, 1985, an unavailable name.

In plane view, the arrangement in an irregular network pattern allies these burrows to *Protopaleodictyon submontanum* (Azpeita Moros, 1933), as shown by Książkewicz (1977) from the Polish Carpathian flysch. *P. submonta-*

num is generally considered a graphoglyptid, being an open burrow system made in mud (Seilacher 1977, as *Megagrapton irregulare* Książkewicz, 1968,; Kern 1980, as *Megagrapton submontanum*), owing its preservation to rapid erosion and filling by turbidites. The Mickwitzia specimens are postdepositional, but the interpretation of graphoglyptids is not unequivocal (see Byers 1982), and Książkiewicz (1977) interpreted *Protopaleodictyon submontanum* as postdepositional. McMenamin & Schulte-McMenamin (1990, Fig. 3:10) illustrate a trace from the Lower Cambrian Gog Group of Canada, similar to *Olenichnus*, which was identified as *Protopaleodictyon*. The trace appears to consist of grooves, but unfortunately it is not stated whether the trace is on an upper or lower surface. Seilacher (1977) restricted *Protopaleodictyon* to regularly meandering forms having distinct appendages. He brought *Cylindrites submontanus* to *Megagrapton irregulare*, defining *Megagrapton* as having 'irregular second order meanders of low amplitude, branching and anastomosing to form meshes of irregular size and shape'. No second order meanders have been recognized in the Mickwitzia sandstone specimens; thus the similarity to *Protopaleodictyon* and *Megagrapton* may be superficial. As far as can be judged, the winding pattern of the Mickwitzia sandstone specimens is similar to that in *Helminthopsis irregulare*, which has irregularly meandering, occasionally void tubes (Książkiewicz 1977). However, *Helminthopsis* does not form burrow systems; the alternative that the specimens described above are chance encounters rather than burrow systems is possible but not likely. Furthermore, *Helminthopsis* is not recorded with vertical segments.

Ichnogenus *Palaeophycus* Hall, 1847

The Palaeophycus/Planolites problem. – The distinction between these two ichnogenera of essentially horizontal cylindrical burrows has received much attention (Osgood 1970; Alpert 1975; Pemberton & Frey 1982; Fillion 1989; Fillion & Pickerill 1990a). The now most widely followed scheme uses the presence or absence of burrow lining and whether the fill is inferred to represent biogenic reworking or physical (or chemical) sedimentation (Pemberton & Frey 1982). Thus *Palaeophycus* are open burrows, typically with a wall-lining and a passive fill, and *Planolites* is used for burrows with a structureless fill differing from the host matrix as the result of the activity of a deposit-feeder (Richter 1937; Osgood 1970; Pemberton & Frey 1982; Fillion & Pickerill 1990a). An example of the reasoning underlying this is found in Richter's (1937) original description of *Planolites montanus* Richter, 1937, from the late Carboniferous of Germany. Richter noted that the burrows occurred in a dark fine-grained laminated clastic sediment with organic material, whereas the material in the tunnel has lightly grayish yellow color and a lower content of carbon. The tunnels were strongly horizontal and lacked vertical components. Therefore, and as no beds of sediment similar to that in the tunnels were observed, Richter (1937) ruled out a passive fill from above into an open tunnel. Instead, the fill was explained as material which had been cleaned of carbonaceous material in the gut and back-filled into the burrow. This is deceptively straightforward and in practice meets considerable problems. One of the more difficult problems results from concealed bed-junction preservation, where the trace is passively filled with a sediment later eroded away or not deposited except in the trace (e.g., Hallam 1975). This may be expected to be common in many shallow-water environments, especially those affected by periodic influx of sediment. If the difference between the intruding and the surrounding sediment is large it may be recognized as passively introduced. However, the smaller the difference the easier it may be interpreted as an active fill. As an example may be given burrows identified as *Planolites annularius* Walcott, 1890, by Fillion & Pickerill (1990a). This identification was based on depletion of mica relative to surrounding sediment, thought to reflect a selective action of the producer. The case was taken as an example that microscopic work may be needed to separate active from passive fill. This rather implies that the uncertainties of identification are at an untenable level. The depletion in mica could have been caused by influx of mica-poor but otherwise similar sediment. The activity of deposit feeders affects the sediment they process. Digestive processes may selectively affect some constituents (see Aller 1982). The important implication for the differing nature of fill in *Planolites* and *Palaeophycus* is that it supposedly reflects different ethological motivation (Pemberton & Frey 1982; Fillion & Pickerill 1990a). However, it is common that deposit feeders do not feed and defecate in the same place. Funnel feeders and conveyor-belt feeders typically feed within the sediment and defecate at the sediment surface, though the opposite also occurs, which may cause a biogenic grading (e.g., Myers 1977; Baumfalk 1979). Often there is one or several vertical shafts leading down to a preferred horizon where a horizontal network of burrows are formed (Bromley 1990). Through the vertical shafts oxygenated water enters the burrow and through the same shafts the animal also disposes of wastes. The open system of the Recent polychaete *Heteromastus filiformis* does not in itself show evidence of deposit feeding, and passive fill would result in a structure classifiable as *Palaeophycus*, especially as its full geometry would rarely be discernable. Modern analogues of *Planolites* are virtually lacking (Pemberton & Frey 1982), contrasting with the numerous reports of this structure in the sedimentary record. This could be explained by diagenetic enhancement, but it also calls for caution. To establish whether the fill resulted from passive sedimentation or biogenic reworking in many situations requires specimen-by-specimen examination and even so would remain conjectural.

No definitive *Planolites* burrows have been found in the Mickwitzia sandstone. A number of specimens, especially those named *Palaeophycus tubularis* small form, are very similar to numerous reports of *Planolites montanus*.

Palaeophycus imbricatus (Torell, 1870)

Figs. 8D, 45, 46, 47B, D, 48B

Synonymy. – □1870 *Halopoa imbricata* n.sp. – Torell, pp. 7–8. □1965 *Halopoa imbricata* Torell 1870 – Martinsson, pp. 219–221, Fig. 29–32. □?1878 *Trichophycus sulcatum* – Miller & Dyer, pp. 4–5, Pl. 4:4. □?1982 *Palaeophycus sulcatus* (Miller & Dyer, 1878) – Pemberton & Frey, pp. 863–864, Pl. 2:2, 3 (see this reference for further synonyms of *Trichophycus sulcatum*).

Lectotype. – Designated by Martinsson (1965, Fig. 29). Labeled SGU Type 5351.

Material. – Common throughout the Mickwitzia sandstone. Especially well-preserved specimens in laminated fine sand- and siltstones of interval B. Figured slabs, SGU Type 5351; RM X231, 3228, 3316, 3319, 3321. Additional material in collections at RM and SGU.

Diagnosis. – Straight to gently winding, mainly horizontal burrows; basically cylindrical, though often with central furrow, and rarely with a crude spreite. Lining generally prominent. Preserved in epirelief, endorelief, exoreliefs or hyporelief. Surface typically with longitudinally or obliquely directed, irregular coarse ridges and wrinkles. Full geometry not recognized, but inclined and vertical segments imply wide tunnel systems.

Description. – Straight to gently winding burrows, mainly horizontal, but inclined and vertical segments occur. Preserved in epirelief, endorelief and hyporelief. Cross-section irregularly cylindrical to vertically extended in form of a crude vertical spreite. Mostly found on tops of rippled beds, where they are preserved as positive epireliefs, formed as endogenic (often intergenic) burrows or full reliefs (Figs. 45, 46). No preferred orientation to ripple morphology. Often the burrow follows the sediment surface for a short distance only, abruptly appearing from and disappearing back into the sediment (Figs. 45, 47D); when penetration is at right angle to sediment surface the burrow is visible as a round knob (Fig. 47D). Occasionally burrows may be traced for longer distances; one specimen propagates in a nearly straight course at about right angle to parallel ripple crests for about 40 cm (Fig. 46A). Its highest relief is found in the ripple troughs (2 mm) while it disappears within ripple tops. Density of burrows may be high and crossings are common and may appear like

Fig. 45. Palaeophycus imbricatus. Specimen designated as lectotype for *Halopoa imbricata* by Martinsson (1965). Scale bar 10 mm. SGU Type 5351.

branches but no true branching has been found. No specimens have been found on soles of sandstones thicker than 3 cm. In inclined specimens there may be an upper core of coarser-grained material, which is surrounded by or lying on top of concavo-convex packs of sediment, the diameter of which increases away from the core (Fig. 48B). Wall often multilayered, best seen on ruptured specimens. Longitudinal sections occasionally show horizontal or low-angle imbricate packing of sand separated by finer-grained partings. Burrow walls ornamented with distinct wrinkled ridges, mainly longitudinally directed; usually traceable for only short distance due to interweav-

ing habit (Fig. 46B). Tension faulting in association with burrow common (Fig. 8D).

HALOPOA. – Hitherto, the only detailed description of *Halopoa* Torell, 1870, is that of Martinsson (1965), which includes the first photographs, and the designation of a lectotype. Martinsson (1965) synonymized Torell's two species of *Halopoa*, *H. imbricata* and *H. composita* as well as *Scotolithus mirabilis* Linnarsson, 1871. Nathorst (1881b, pp. 30, 52) considered *H. imbricata* to have been formed by a crustacean, and compared it with swimming trails of *Corophium*, whereas *H. composita* was thought to

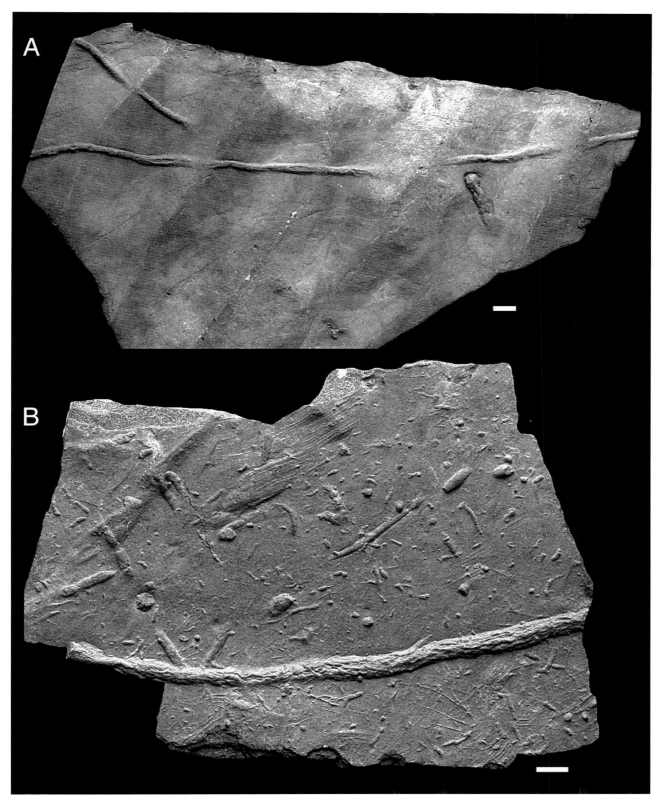

Fig. 46. Palaeophycus imbricatus. □A. Specimen following rippled surface. Scale bar 10 mm. Hjälmsäter, basal part of interval B. RM X3228 □B. Specimen on lower surface of laminated fine sandstone, with wrinkled surface caused by tension. Also seen are '*Eophyton*'-type tool marks and fine ridges possibly caused by claws of current-swept trilobites. Scale bar 10 mm. Lugnås. Interval B. RM X231.

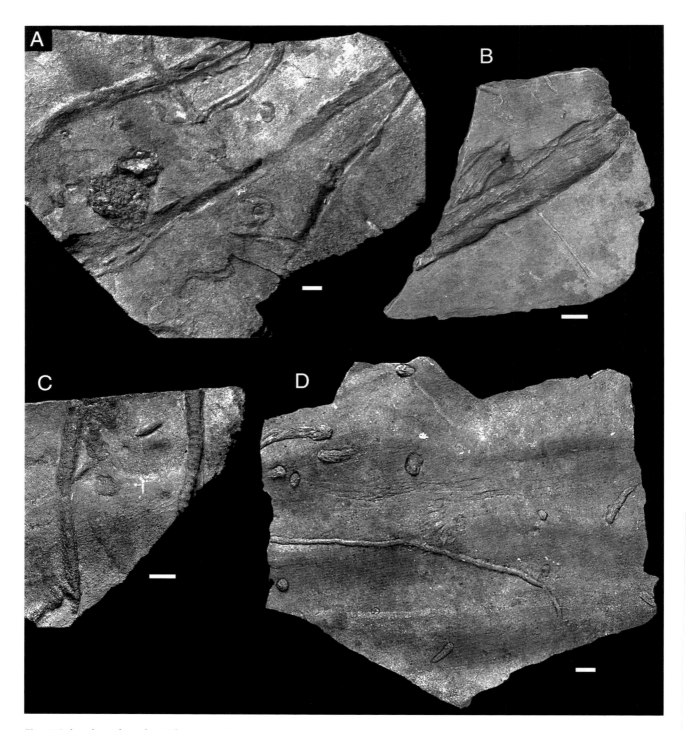

Fig. 47. *Palaeophycus* from the Mickwitzia sandstone. Scale bars 10 mm. □A. *Palaeophycus tubularis.* Probably collapsed specimen. Possible bilobate development along parts of specimen below center could make the ichnogeneric assignment doubtful. Stora Stolan, Billingen. RM X3316. □B. *Palaeophycus imbricatus* in sole view. Stora Rud, Lugnås. RM X3317. □C. *Palaeophycus tubularis.* The specimen on the right side shows transverse ridges of uncertain origin. Stora Stolan, Billingen. RM X3318. □D. *Palaeophycus imbricatus* on rippled surface. Note specimens cutting bedding at nearly vertical angles. Lugnås. RM X3319 .

be made by a worm. Torell (1870) described *H. composita* as having an ear of grains (*Halopoa* was interpreted as a plant) consisting of unilaterally positioned, elongated to spear-shaped grains, and a long and straight stem with

barely visible leaves. This description agrees well with a specimen labelled as type material in the collection of SGU, with number SGU Type 5353 (Figs. 50, 51). Burrows matching the description of ears of grains are seen

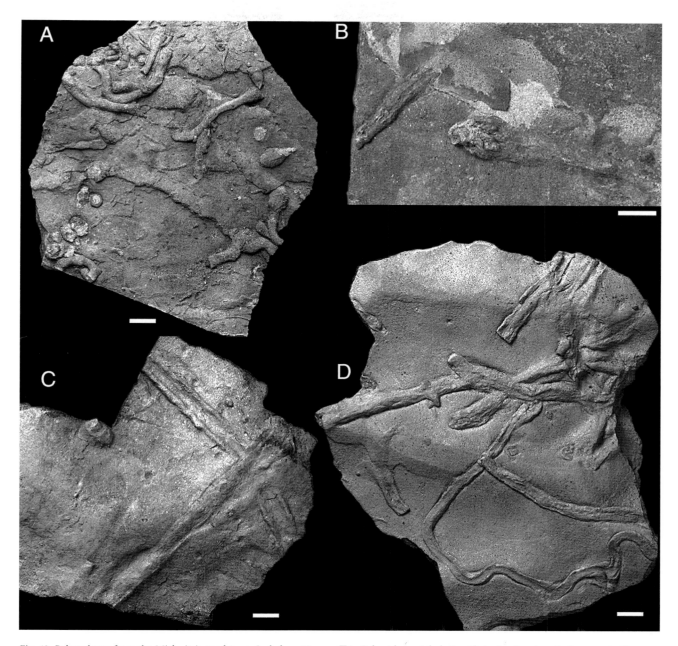

Fig. 48. Palaeophycus from the Mickwitzia sandstone. Scale bars 10 mm. □A. *Palaeophycus tubularis* with inclined and vertical segments. Top part expanded and usually with a surrounding collar. Expanded upper part of burrow may have longitudinal ridges, Scale bar 10 mm. Lugnås. RM X3324. □B. *Palaeophycus imbricatus.* Upper surface view. Note top portion of a burrow with a central coarse-grained fill and a wrinkled cover. Stora Rud, Lugnås. RM X3321 □C. *Palaeophycus tubularis.* Specimen preserved on upper surface of rippled bed, exhibiting burrow collapse. Lugnås. RM X3322. □D. *Palaeophycus tubularis* on rippled bed. Burrow collapse common. Specimen in lower part of picture has winding habit similar to that of *Helminthopsis.* Lugnås. SGU 8595.

just below the center of Fig. 51B. This shows that *H. composita* was designated for burrows of a different type than *H. imbricata*, and since the long stems represent *H. imbricata* the type material is inhomogeneous. The pattern described as ears of grain are here assigned to *Phycodes* cf. *curvipalmatum* Pollard, 1981. Linnarsson's (1871) description of *Scotolithus mirabilis* stressed the radiating

arrangement of the burrows; it is not assignable to *Halopoa*. Therefore, in the following discussion only *Halopoa imbricata* is considered. Torell (1870) described *H. imbricata* as a plant, identifying stem, leaves and spikes. It is not easy to identify these features on the lectotype specimen, though the leaves seem to have been identified with imbricated patterns on the burrow-wall. An expanded top with spikes was given as more than twice the size of the stem and seems to correspond to multi-walled areas, probably close to the burrow opening (see below).

Martinsson (1965) compared *Halopoa* with *Gyrochorte* and thought it possible that the difference between the two might 'turn out to be due to difference in lithology and preservation'. Though there exists similarity in plan view, the internal structure is different. *Gyrochorte comosa* Heer, 1865, is a trace less than 10 mm wide, preserved in positive epirelief as two ridges with obliquely aligned pads, and in hyporelief as biserial smooth grooves. Actual connection between epirelief and hyporelief were observed by Weiss (1941) and have been described in detail by Heinberg (1973). The mode of formation for this trace is generally thought to be digging by a worm-like organism in a progressive oblique stance through the sediment (Weiss 1941; Seilacher 1955b; Heinberg 1973), whereby sediment becomes pushed up in pads above the sediment surface. *Halopoa imbricata* differs from *Gyrochorte* in being originally unilobed and having little oblique additive digging.

In its ornamentation and habit *Halopoa imbricata* is very similar to burrows reported as *Fucusopsis* Palibin *in* Vassoevich, 1932, and *Palaeophycus sulcatus* (Miller & Dyer, 1878). Hakes (1976, p. 27) compared *Halopoa* to specimens of *Fucusopsis* from upper Pennsylvanian of Kansas (identified as *Palaeophycus sulcatus* by Pemberton & Frey 1982, p. 863). He considered surface sculpture to be useful in differentiation, stating that the ridges in *Halopoa* are 'not as delicately sculptured and are proportionally much shorter than those described here'. The sculpture on *Fucusopsis* has been interpreted as resulting from the digging activity of the animal (Seilacher 1959a). According to Osgood (1970, p. 381), '[t]he pressure of the body produced a series of faults in the basal lamina of the host rock, although occasionally the organism broke through the base of the bed to give rise to longitudinal striae'. Obvious faulting is found also in the Mickwitzia specimens, and it is clear that also the longitudinal striae formed through deflected and ruptured laminae (Fig. 20B$_3$). Examination of Paleogene material of probable *Palaeophycus sulcatus* from Punta del Carnero, southern Spain, shows that also in these, the longitudinal ornamentation is caused by deflected sediment. Judging from the literature of *Fucusopsis* and *Palaeophycus sulcatus* this could be the general case. The diagnostic value of this sculpture would thus be limited as it is reflective of properties of the sediment.

Palaeophycus, was diagnosed by Fillion & Pickerill (1984) as follows: 'Branched or unbranched straight to slightly curved to slightly undulose or flexuos, smooth or ornamented, lined essentially cylindrical, predominantly horizontal burrows of variable diameter; infilling typically structureless, of same lithology as host rock.' Reports of *Palaeophycus* generally do not include spreite-structures,

but the morphology of specimens reported as *Fucusopsis* and *Palaeophycus sulcatus* is identical to that of *Halopoa*, and a crude vertical stacking is seen on specimens reported by Fillion & Pickerill (1990a, Pl. 11:2, 5). *Halopoa imbricata* is therefore transferred to *Palaeophycus*, and it may be the same type of burrow as *Palaeophycus sulcatus*. The interpretation of the origin of the burrow sculpture makes the concept of this ichnospecies doubtful, except as an indicator of sediment properties.

It is difficult to evaluate the few reported occurrences of *Halopoa* beside those from the Lower Cambrian of Västergötland and Middle Cambrian of Öland. *Halopoa* was reported from the Talsy horizon in Estonia (Palij *et al.* 1983, Pl. 67:1–2). As judged from the illustrations and from my own observations on material in the Kopli quarry, Tallinn, these are *Palaeophycus imbricatus*. Poulsen (1967, p. 37, Fig. 5) reported *Halopoa* cf. *imbricata* from siltstone beds of the 'Green Shale' on Bornholm. However, Clausen & Vilhjalmson (1986) give no mention of *Halopoa* in a study of trace fossils from the same level and area, though they report *Palaeophycus tubularis*. The available information prevents judgement on *Halopoa* from the early Cambrian Tal Formation, Lesser Himalaya, India (Singh & Rai 1983), and Jurassic specimens from Kutch, India (Badve & Ghare 1978); for the latter occurrence a new species, *Halopoa indica,* was erected.

Interpretation. – *Palaeophycus imbricatus* in the Mickwitzia sandstone is interpreted as a dwelling or feeding–dwelling structure that may have been wide U- or L-shaped burrows of which some may have been joined to a common entrance. These were traces burrowed within mud as well as sand; crude vertical repetition of the burrow is interpreted as response to influx of sediment, though some of the vertical displacement may represent search for food. Multilayered walls and downward increasing width of laminae in inclined segments is suggestive of sediment being pressed into the walls or cleaned out from the burrow and included in the burrow lining, a behavior seen today among various burrowers (Reineck 1958). Vertical or inclined shafts leading down to preferred levels were probably open and repeatedly used, as shown by diverging burrows. On ridges preserved in hyporelief there may still be found drag-marks plastered onto the burrow, showing that sediment was pressed out during burrowing, and that the animal itself did not break through the silt–mud boundary. The most delicately wrinkled specimens occur with fine-laminated siltstones, and it is suggested that this sculpture formed entirely by fracturing of cohesive sediments as laminae were deflected and ruptured. In sectioned specimens, down-bent laminae can actually be followed to the lower surface. Differential weathering of the material in the laminae occasionally lead to a structure of separated flakes of

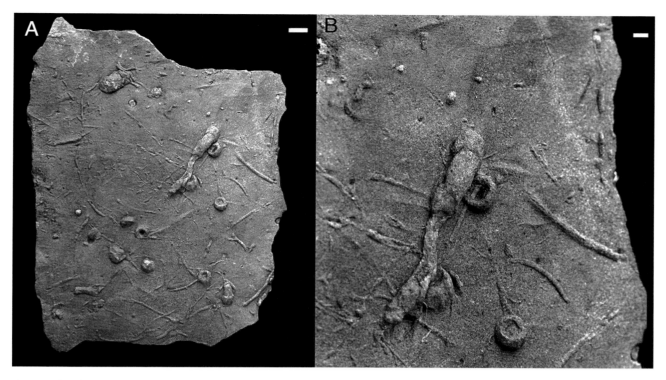

Fig. 49. Palaeophycus tubularis. □A. *Palaeophycus tubularis,* small form, and inclined parts of *Palaeophycus tubularis.* Scale bar 10 mm. Lugnås, RM X3323. □B. Close-up of A. Note wall lining, collapse of large form and the upper capping collar. Scale bar 2.5 mm.

sandstone. Specimens with imbricated sides reflect tension in the sediment. *Palaeophycus imbricatus* appears to have been made by an organism pushing itself through, rather than excavating, the sediment. The resulting morphology will largely reflect the condition of the sediment.

Palaeophycus tubularis Hall, 1847

Figs. 47A, C, 48A, C, D, 49

Material. – Figured specimens on slabs, RM X3316, 3318, 3320, 3322; SGU 8595. Additional material at RM and SGU. Found at Kinnekulle and Lugnås.

Description. – Straight to gently curving, essentially horizontal burrows with smooth walls that may be collapsed (Fig. 48C). In inclined or vertical parts the uppermost portion may be expanded into a collar (Fig. 48A). Lining may be prominent. Diameter of burrows 3–12 mm.

Discussion. – These are open dwelling or dwelling–feeding burrows preserved by infalling sediment (Fig. 20B$_1$). Expanded upper parts may have been formed by the animals activity or by scouring before or during sedimentation. This upper part has probably been identified in part as *Micrapium erectum.*

Palaeophycus tubularis Hall, 1847, small form

Figs. 40B, 49

Material. – Specimens examined on 15 slabs; figured RM X3310, 3323–4. Found at Kinnekulle and Lugnås. Interval B.

Description. – Straight to gently winding to contorted, essentially horizontal burrows often occurring in dense congregations (Fig. 49). Cross-section circular to flattened in vertical direction. Preserved as exoreliefs and hyporeliefs. Found in intervals of alternating thin silt–sandstone and clayey layers. Exoreliefs with a filling similar to or cleaner and coarser than overlying sand–siltstone. Surface of burrow without obvious ornamentation, except for grainy surface reflecting coarse material in burrow. Some lower surfaces have abundant short segments and small knobs of the same dimension as the horizontal segments. Several burrows may radiate from a small area, and intensive burrowing may result in heaps where individual burrows are poorly seen. In places branching appears to occur, though it is not possible to decide whether this is true branching. Diameter of burrows 1–2 mm.

Discussion. – These burrows fall within the morphological range of *Planolites montanus* as given by Pemberton & Frey (1982) and later emended by Fillion & Pickerill (1990a), though as argued above assignment is made to *Palaeophycus*. The smooth walls and occasional gentle collapse is consistent with *Palaeophycus tubularis*. Local concentrations of burrows and crude radiating arrangement imply a burrow system rather than isolated tunnels. The varying morphology may depend on relative abundance of horizontal, inclined, vertical and possibly undulose segments but could also depend on the level at which the burrows are preserved. Occurrences with abundant vertical knobs are similar to *Planolites punctatus* Roniewicz & Piénkowski, 1977. I concur with Pemberton & Frey (1982) that this type of burrow should be placed in *Planolites montanus* (for different opinion see Piénkowski & Westwalewicz-Mogilska 1986). Probably related to this burrow in the Mickwitzia sandstone is *Olenichnus* isp., which differs in having larger, more winding burrows, and a different mode of preservation.

Ichnogenus Phycodes Richter, 1850

Phycodes cf. *curvipalmatum* Pollard, 1981
Figs. 40B?, 55A

Material. Figured slabs, SGU Type 5352, RM X3310. This burrow is common in interval B (about 3.5 m above basement) in mines at Älerud, Lugnås.

Description. – Closely packed burrows, parallel at proximal end, then gently diverging to give a fasciculate appearance, or with oblique, closely spaced short branches (Fig. 51A). Burrows more or less horizontal in proximal part, bending sharply upwards in the distal portion. Number of branches low, about five. Diameter of burrows 3.5–6.5 mm. Surface of burrow smooth or wrinkled.

Discussion. – These burrows show characters intermediate between *Phycodes*, with branches originating from a

Fig. 50. Type slab of *Halopoa composita*, a species that is too heterogenous to warrant retention. Long, more or less straight specimens can be assigned to *Palaeophycus imbricatus*. Others show branching similar to that of *Phycodes palmatus*, *Phycodes* cf. *curvipalmatum* and *Treptichnus pedum*. Scale bar 10 mm. Lugnås. SGU Type 5352.

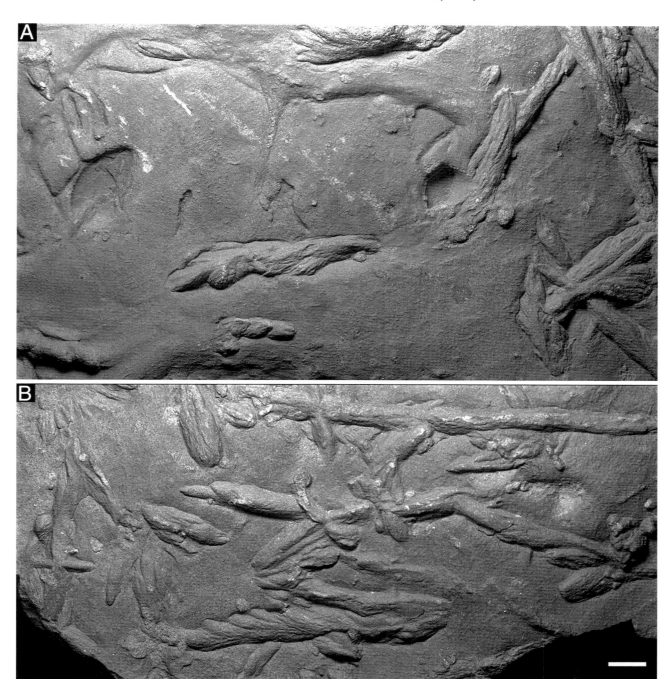

Fig. 51. Details from slab SGU Type 5353., Lugnås, interval B. Scale bar 10 mm. □A. Specimens of *Phycodes* cf *curvipalmatum* (center) and *Phycodes palmatus* (right). □B. Several types of burrows, including *Phycodes* cf. *curvipalmatum.*

more or less common point, and *Treptichnus*, with advancing segments. This is also met with in *Phycodes curvipalmatum* Pollard, 1981, described from the Triassic of Cheshire, England (Pollard 1981). It consists of compressed burrows that branch in a dichotomous or palmate manner. Some branches are strongly recurved, forming a right angle to the main burrow. From Pollard's (1981) description it is not clear whether the branches also curve

up, though this seems likely from the pointed ends of the branches in figured specimens (e.g., Pollard 1981, Fig. 7). Eagar *et al.* (1985) reported *Phycodes* cf. *P. curvipalmatum* to have short radially inflated lobes curving both to the sides and upwards. Consequently, the specimens from the Mickwitzia sandstone are tentatively assigned to *Phycodes curvipalmatum.* Eagar *et al.* (1985) reported their burrows to be highly variable; this is also the case for the specimens described above. This form is also similar to reports of *Phycodes pedum* Seilacher, 1955, though the holotype of that species have burrows advancing in the manner of *Treptichnus*. However, *Phycodes curvipalmatum* and *Treptichnus pedum* may be transitional forms.

There is also similarity with specimens reported as *Phycodes palmatus* (Hall, 1852). A reconstruction of *P. palmatus* by Seilacher (1955b, Fig. 5:14) is similar to the specimen illustrated here in Fig. 40B. However, Fillion & Pickerill (1990a, p. 47) diagnosed *Phycodes palmatus* (Hall, 1852) as '*Phycodes* exhibiting a few thick and rounded branches, branching horizontally in a palmate or digitate form from nearly the same point'. According to Pollard (1981), *P. curvipalmatum* differs from *Phycodes palmatus* in having shorter, more curved branches.

Phycodes palmatus (Hall, 1852)

Figs. 40B?, 50, 51B

Material. – Common on beds about 3.5 m above basement in mines at Älerud, Lugnås. Specimens in the collections of RM and Jan Johansson, Sköllersta. Figured specimens on slab SGU Type 5353.

Description. – Essentially horizontal, endogenic burrow systems with a palmate appearance in hyporelief preservation (Fig. 50). Proximal part not observed; distally branching into a few to about 20 straight or gently curved burrows. Diameter of individual burrows 3–6 mm.

Discussion. These specimens are assigned to *Phycodes palmatus* because of the small number of burrows, their relatively large size and the branching from more or less the same point. *Phycodes palmatus* is found together with *Scotolithus mirabilis* and *Phycodes* cf. *curvipalmatum*. These taxa are found on the same bedding planes and probably form a morphologically continuous series. *Phycodes palmatus* is widely found in Lower Cambrian strata, including; the Green Shales on Bornholm (Clausen & Vilhjalmson 1986), the Torneträsk Formation, northern Sweden (personal observation), the Norretorp Formation, southern Sweden (personal observation).

Ichnogenus *Rhizocorallium* Zenker 1836

Rhizocorallium jenense Zenker, 1836

Figs. 6A–B, 40A, 52, 65A?

Material. – Hundreds of specimens on bedding planes 600 m south of Hällekis, about 6 m above the base of the conglomerate. Specimens in loose block also from Lugnås. Figured specimens RM X3309, 3326.

Description. – Horizontal to inclined, curved to straight protrusive spreiten, flanked by a U-shaped, variably preserved, occasionally vertically retrusive, marginal burrow (Fig. 52). Width of spreite 40–53 mm, except one slab with a possible *R. jenense* that is about 14 mm wide (Fig. 65A). Spreite consists of closely stacked arcuate ridges that may differ from the surrounding sediment by higher content of silty–clayey material and/or higher content of coarse material. The ridges are convex, following outline of U-tube. In vertical profile the laminae of the spreite are typically inclined backwards and may be faintly concave to the anterior direction. Marginal burrow preserved as a furrow or sand-filled burrow that may have a retrusive or possibly combined retrusive–protrusive spreite. Marginal burrow may show lateral and longitudinal displacement. Width of marginal burrow about 5–10 mm. Width of spreite 40–55 mm. Length of longest fragment 140 mm. Preserved in negative or positive epirelief. Most specimens found in medium-grained sandstone.

Discussion. – The more or less horizontal expression of the spreite and the marginal flanking U-shaped burrow place these trace fossils in *Rhizocorallium*. Fürsich (1974b) recognized three ichnospecies of *Rhizocorallium* based upon general shape and orientation of the spreite. *R. jenense* includes short and straight forms generally oriented obliquely to the bedding plane. Long, mainly horizontal specimens, sinuous, bifurcating or planispiral, are included in *R. irregulare* Mayer, 1954, and trochospirals in *R. uliarense* Firtion, 1958. Though followed by most subsequent workers, the proposal was criticized by Hecker (1980, 1983), who argued that *Rhizocorallium* as used by Fürsich is too wide a grouping and that the material upon which the type species, *R. jenense* Zenker, 1836, was described consists of specimens that have a vertical to inclined orientation relative to bedding and strong scratch-marks. For strictly horizontal specimens with strongly parallel limbs and smooth marginal burrows, Hecker (1980) erected the new ichnogenus *Ilmenichnus*, with type species *Rhizocorallium devonicum* Hecker, 1930. In Hecker's (1980, p. 19) opinion, *R. irregulare* Mayer, 1954, was based on insufficient and uncharacteristic material. I concur with Fürsich (1974b) in regarding the presence or absence of scratch marks as inappropriate in spreite-burrow taxonomy. However, following the more detailed picture of *R. devonicum* Hecker, 1930, emerging from Hecker's later works (1980, 1983) it seems that *R. devonicum* more closely matches the diagnostic features of *R. irregulare* than it does *R. jenense*, as was tentatively suggested by Fürsich (1974b).

In possessing a short horizontal to inclined spreite with some retrusive development of the marginal burrow, specimens from the Mickwitzia sandstone are referable to *R. jenense*. In frequently having a curved outline, and with some specimens having horizontal sections of several dec-

Fig. 52. □A. *Rhizocorallium jenense* preserved on top of rippled, medium-grained sandstone. Scale bar 10 mm. Hällekis. RM X3326.

imetres' length, they intergrade with *R. irregulare* or, perhaps more properly, *R. devonicum*. The discontinuity of the marginal burrow shows that the spreite was not formed through a smooth forward lengthening. Vertically retrusive marginal tunnels probably represent response to sediment entering the burrow. Laterally displaced and inclined parts of the marginal tube show little vertical displacement and may represent substantial displacement of the entire burrow in a longitudinal direction. This may have been caused by limits posed by circulation of oxygenated water in a tube exceeding a certain length. *R. jenense* is interpreted as a dwelling or dwelling–feeding burrow (Seilacher 1967; Fürsich 1974b).

Rhizocorallium jenense is typically found in high-energy shallow marine conditions (e.g., Seilacher 1967), though it has also been found in non-marine beds (Fürsich & Mayr 1981). In the Mickwitzia sandstone, rich occurrences of *Rhizocorallium* are found in intervals A and C, in beds with coarse sand and mud clasts.

R. jenense has been reported from the following early Cambrian beds: the Neobolus Beds of the Salt Range, Pakistan (Seilacher 1955b); the Broens Odde Member of the Laeså Formation on Bornholm, Denmark (Clausen & Vilhjalmsson 1986); the Ocieseki Sandstone Formation of the Holy Cross Mountains, Poland (Orłowski 1989). *R.* cf. *jenense* has been reported from the Pestrotsvet Formation of Aldan, Siberia, Russia (Fedonkin 1981). *Rhizocorallium*

sp. has been reported from the upper part of the Doulbasgaissa Formation in Finnmark, Norway (Banks 1970); the Lükati Formation of Estonia (Palij *et al.* 1983) and from cores of southeastern Poland in the Mazowsze Formation (Fedonkin 1981). The Mazowsze Formation correlates with the Lontova horizon; other occurrences from the Baltoscandian area correlate with the Talsy horizon or higher. The Siberian occurrence belongs to the Tommotian Stage.

Ichnogenus *Rosselia* Dahmer, 1937

Rosselia socialis Dahmer, 1937

Fig. 53C, E; 65C

Material. – Six slabs; figured RM X3329, 3331, 3341. Lugnås, Kinnekulle.

Description. – Irregularly conical to bulbous burrows with a flat to gently convex top surfaces possessing concentric, irregular, lamination and a central to sub-central, vertical to inclined tube (Fig. 53C, E). Burrows are directed from a vertical to a near horizontal stance. In vertical sections, burrows are sharply separated from the surrounding sediment, from which they differ by higher content of clayey material and also a higher number of coarse sand grains.

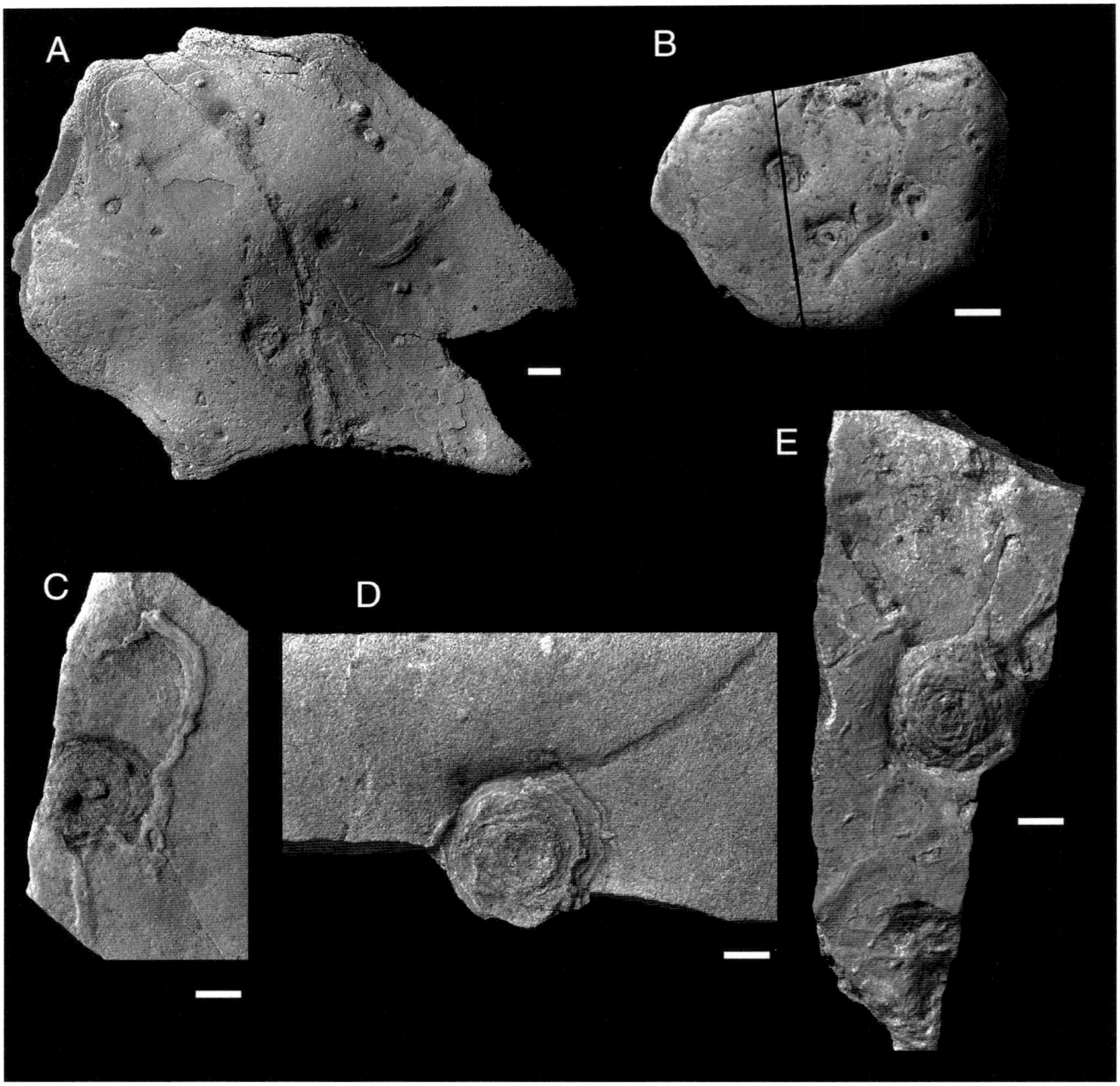

Fig. 53. Rosselia and other funnel-topped burrows. Scale bars 10 mm. □A. Upper surface of slab with several knobs representing sections through vertical burrows. Trolmen. RM 3327. □B. Top view of vertical tubes with concentric lamination. Trolmen. RM X3328. □C. *Rosselia socialis.* Top view. Fine concentric lamination around a subcentral plug. Trolmen. RM X3329. □D. Top view of bluntly conical funnel, similar to *Conichnus.* Trolmen. RM X3330. □ E. *Rosselia socialis,* top view of bulbous burrow with concentric ornamentation. Hjälmsäter. RM X3331.

Continuation of concentric laminae poorly visible in sectioned specimens. In sectioned specimens the sand-filled tube bends towards a horizontal orientation below the funnel. In one specimen, flatly cylindrical structures radiate out from base of burrow onto lower bedding surface. Height up to 30 mm, width 25–47 mm. Dimension of central tube about 3 mm.

Highly inclined specimens have tube positioned high on arrow-head-shaped crudely laminated area (Fig. 65C) These inclined specimens are often connected to *Palaeophycus imbricatus.*

Discussion. – The burrows show similarity to *Rosselia socialis* as described by Dahmer (1937) from the Devonian

Taunusquartzit, Germany. *Rosselia* has usually been interpreted as having been formed by the rotatory movement of a deposit feeder, or by infalling material being pressed to the walls of an open dwelling burrow. The latter activity was termed 'Räum-Auskleiding' by Reineck (1958, p. 7), who gave examples of such structures produced by *Nereis diversicolor* under conditions of slow sedimentation. In inclined burrows the infilling material is mostly pressed to the floor (Reineck 1958). This is analogous with inclined specimens of *Rosselia* described above, with the upper tube possibly caused by heavy influx of sediment leading to abandonment or death of the animal. Rindsberg & Gastaldo (1990) reported brackish-water specimens of Holocene *Rosselia* from Alabama to have formed as the reinforced top of the crustacean burrow *Thalassinoides* in soft to soupy sediments.

Three species are currently recognized in *Rosselia*. Of these *R. socialis* has concentric lamination surrounding a tube, while *R. rotatus* McCarthy, 1979, has crescentic back-fill structures within the funnel (McCarthy 1979). *R. chonoides* Howard & Frey, 1984, has helicoidal swirls of reworked sediment (Howard & Frey 1984). Though the internal lamination is poorly visible, the upper concentric lamination and the general shape of the burrow suggest that the specimens from the Mickwitzia sandstone are *R. socialis*. The interpretation favored here of *R. socialis* promotes a review of the generally accepted mode of formation of *R. socialis* and raises the issue whether the two other species currently recognized should be assigned to other ichnogenera. Uchman & Krenmayer (1995) consider *R. rotatus* to fall within the morphological range of *R. socialis* and *R. chonoides* to be secondarily reworked *R. socialis* or to be assignable to a different ichnogenus.

Ichnogenus *Rusophycus* Hall, 1852

Rusophycus dispar Linnarsson, 1869

Figs. 54–57

Synonymy. – □*pars* 1869 *Rhysophycus dispar* n.sp. – Linnarsson, pp. 353–356. □*pars* 1870 *Cruziana dispar* Lins. sp. – Torell, p. 6. □*pars* 1871 *Cruziana dispar* Linnarsson – Linnarsson, pp. 14–16, Figs. 17, 18, *non* Fig. 19 (= *Cruziana rusoformis*). □*pars* 1968 *Cruziana dispar* Linnarsson, 1871 – Bergström, pp. 499–500, Fig. 7 (? = *Cruziana rusoformis*). □*pars* 1970 *Cruziana rusoformis* ichnosp. nov. – Orłowski, Radwański & Roniewicz, pp. 348, 356, Pl. 1a, *non* Pl. 1b, 2b (= *Cruziana regularis*), *non* Pl. 1c (= *Rusophycus radwanskii* Alpert, 1976). □*pars* 1970 *Cruziana dispar* Linnarsson – Seilacher, pp. 453, 457, Figs. 2c, 5c, ?1c. □*pars* 1973 *Cruziana dispar* Linnarsson, 1869 – Bergström, pp. 52–55, Fig. 16, Pl. 5:9, ?Pl. 5:12, *non* Pl. 5:10, 13–14 (= *Cruziana rusoformis*). □*pars* 1990 *Rusophycus dispar* Linnarsson, 1869 – Jensen, p. 31, Figs. 1, 4a, 9, 10, *non* Figs. 2, 3, 4b, 8 (= *Cruziana rusoformis*). □?*pars* 1992 *Cruziana dispar* – Orłowski, p. 19, Fig. 3. □? 1967 *Cruziana dispar* Linnarsson – Poulsen, p. 40, Fig. 6. □? 1983 *Rusophycus* sp. A – Palij, Posti & Fedonkin, p. 92, Pl. 67:3. □*non* 1973 *Cruziana* cf. *dispar* Linnarsson 1889 – Bandel, pp. 169–170, Pl. 38:3. □*non* 1976 *Rusophycus* cf. *R. dispar* – Alpert, p. 229, Pl. 2:1, 3, 5. □*non* 1986 *Cruziana dispar elongata* subsp. n – Wang & Zhang, p. 113, Pl. 1:13 (nomen nudum). □*non* 1988 *Cruziana* cf. *C. dispar* Linnarsson, 1869 – Bergström & Peel, pp. 50–52, Fig. 8. □*non* 1990 *Rusophycus dispar* Linnarsson, 1869 – Pickerill & Peel, pp. 30–31, Fig. 12c, d (? = *Rusophycus latus* Webby, 1983).

Material. – *Rusophycus dispar* is one of the most common trace fossils in the Mickwitzia sandstone. Especially rich museum collections in RM. Figured slabs, PMU Vg 992; RM X695, 3332–3; SGU 8596–7.

Description. – Bilobate burrows with heart-shaped or oval outline (Figs. 54–57). Width and length about equal in most specimens; in deep burrows length is greater than width (Fig. 56). Lobes often strongly convex, with steep sides and a pronounced median furrow. Median furrow deepens and widens in the wider, anterior end of the trace, often entirely separating the two lobes (Figs. 54A, 56A, 57). The lobes are set with ridges that are transverse at the central, deepest part of the burrow, while in the anterior and posterior parts they are proverse and retroverse, respectively (e.g., Fig. 55B). The proverse and transverse ridges are mostly straight and deviate little from vertical inclination, while the retroverse ridges are straight or bend posteriorly and are inclined backwards. The ridges are coarse and mostly paired; occasionally a claw pattern indicating four, possibly five claws is seen on the proverse markings (Fig. 57B). The transition between proverse and retroverse digging is often distinct and may be marked by the transverse and proverse diggings being more pronounced (Figs. 55B). In the area of proverse digging the ridges are typically irregularly spaced. Rare deep forms may show carapace imprints on the burrow sides (Figs. 55B–C, 56; described in greater detail below). Alternation in digging positions occurs (Fig. 55A), and the trace may be clustered and integrade with *Cruziana rusoformis*.

Discussion. – Linnarsson (1869, 1871) stressed bidirectional scratch marks and an anteriorly widening median divide as characteristics of *Rusophycus dispar*. I follow these criteria and do not include in *R. dispar* forms with a possibly similar scratch pattern, but without bidirectional digging (Alpert 1976b; Bergström & Peel 1988; Pickerill & Peel 1990).

In his monographic treatment of *Cruziana* and *Rusophycus*, Seilacher (1970, p. 457) erected a 'dispar group' covering forms having 'several but unequal, sharp

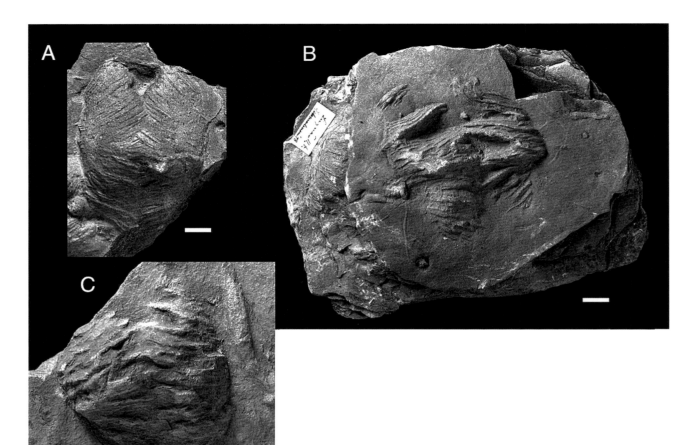

Fig. 54. Rusophycus dispar from the Mickwitzia sandstone. Scale bars 10 mm. □A. Specimen with distinct anterior cleft, separating the two lobes of scratch marks. Lugnås. SGU 8596. □B. Specimen with well-developed, and partly isolated scratch marks in the anterior part. Lugnås. SGU 8597. □C. Specimen with poorly developed retroverse scratch marks. Hällekis. RM X3332.

scratches within each endopodal marking' and with an obvious divide between proverse and retroverse markings. In this group Seilacher (1970) included *Rusophycus dispar*, *R. ramellensis* Legg, 1985 (rusophyciform *Cruziana barbata* of Seilacher, 1970), and *R. grenvillense* Billings, 1862. *Rusophycus ramellensis* Legg, 1985, differs from *R. dispar* in having a more marked depth difference between proverse and retroverse legmarks, with often only weakly developed or absent retroverse markings (Seilacher 1970; Legg 1985). As seen from specimens illustrated by Legg (1985), the ridges in the anterior part of *R. ramellensis* are less strongly proverse, somewhat curved, and often closely follow the burrow outline. Seilacher's (1970, p. 457, Fig. 1) suggestion that *R. dispar* was more deeply cut into the mud than *R. ramellensis* is questionable; shallow traces are by far the more common in *R. dispar*, though museum collections show a bias for deep specimens. Reports of *R. ramellensis* have so far been restricted to the Middle Cambrian, for which it has been used as an index fossil (Seilacher 1970; Yang 1990). *R. grenvillense* Billings, 1892, has an outline similar to deep

R. dispar. Bidirectional scratch marks are not evident on specimens figured by Hofmann (1979, Pl. 10A–D), but they were reported by Seilacher (1970), who also remarked that the endopodal markings of *R. grenvillense* are probably different from those of *R. dispar*. Until these markings are well known, further comparison is not possible. Lower Cambrian specimens described as *Cruziana* cf. *C. dispar* by Bergström & Peel (1988), *Rusophycus dispar* by Pickerill & Peel (1990) and *R.* cf. *dispar* by Alpert (1976b), all lack a clear distinction between proverse and retroverse diggings and bear greater similarity with *Rusophycus latus* or *Cruziana rusoformis*. *R. radwanskii* Alpert, 1976 (formal name for '*Cruziana rusoformis*' Orłowski,

Fig. 55 (opposite page). *Rusophycus dispar* from the Mickwitzia sandstone. □A. Large specimen with two overlapping directions of digging. Scale bar 10 mm. Lugnås. RM X3333. □B. Specimens with impressions of the exoskeleton flanking the burrow. Scale bar 10 mm. Lugnås, PMU Vg 992. □C. Detail of specimen in B., with repeated impressions of genal spines. Scale bar 5 mm.

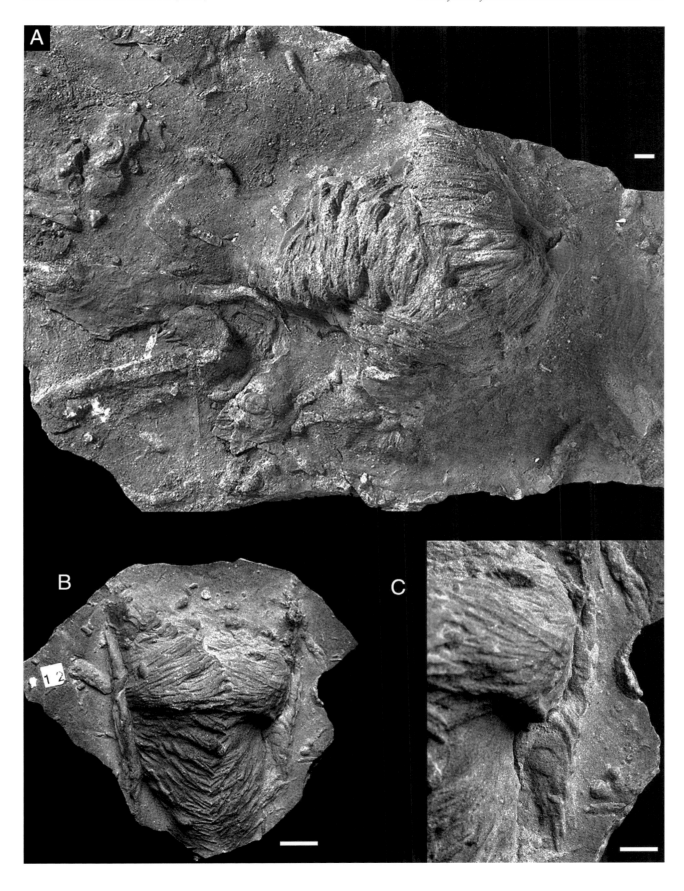

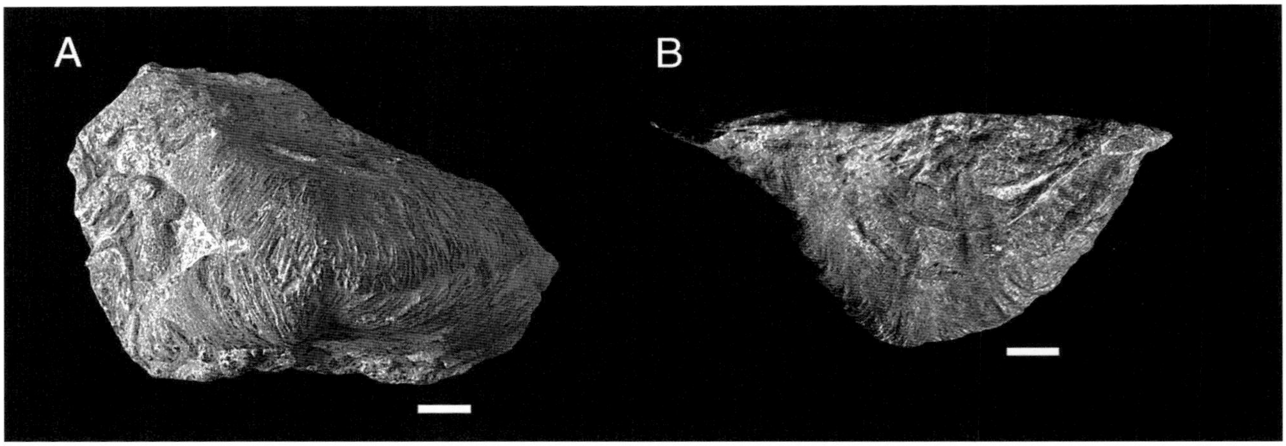

Fig. 56. Deep *Rusophycus dispar*, seen from below (A), and from the side (B). Scale bar 10 mm. Lugnås. RM X695.

Radwański & Roniewicz, 1970 [*partim*]), differs from *R. dispar* by showing strong rotational activity.

The wider end of *Rusophycus dispar* was suggested to correspond to the anterior end of a trilobite by Seilacher (1959b, Fig. 4; 1970, Figs. 1C, 5C). This is supported also by specimens of *R. dispar* with impressions of the exoskeleton. The most well-preserved specimen has two narrow bands bordering and partly truncated by the scratch lobes (Fig. 55B). The right border (as oriented in the figure) has a repeated shallow crescent-shaped pattern with pointed structures directed strongly rearward and slightly outward (Fig. 55C). The crescents are delineated by steep rims in the rear part and become successively somewhat deeper anteriorly. In the anterior part of the trace, the pattern is more variable. These structures are probably impressions of the sides of the cephalon, and particularly of the genal spines. The repetition shows that the trilobite moved forwards in steps and that some retracking occurred. The vertical inclination of the claw impressions shows the legs in the anterior part to have been brought down in a vertical or near-vertical stance during digging. The often strong convexity of the burrows thus depends on the direction in which the legs were brought together and possibly also on the unequal length of the legs, in addition to a backward-bent stance.

The burrow with head-shield impressions suggests that at least the deep burrows were not formed in one single stance. Rather, the animal first dug repeated burrows where the legs were brought together in a backward–medial stance and then made the deeper traces in the anterior end. The trace could thus be seen as a combined *Cruziana* and *Rusophycus*. Though much of the morphologic variation in *R. dispar* can be explained by how deeply the trace impinged into mud, the morphology also depends on the digging posture of the animal. This is most clearly seen in specimens with wide areas of proverse

diggings and weakly developed retroverse diggings; in these the claws in the proverse region were inclined backwards, indicating a head-down digging posture, probably with a strongly bent thorax. The morphologic range seen in *R. dispar* indicates that besides purely anatomic features of the producer, the difference between *R. dispar* and *R. ramellensis* could depend on different preferred digging postures. *R. bonnarensis* Crimes *et al.*, 1977, may represent a further extreme, as it consists solely of moustache-shaped anterior diggings (see Crimes *et al.* 1977).

Much discussion has focused on the originator of *R. dispar* (Bergström 1968, 1973a; Hesselbo 1988; Jensen 1990), mainly concerning the possibility that it was made by *Paleomerus hamiltoni*. Uncertainties whether *Paleomerus hamiltoni* occurs in the Lingulid, rather than Mickwitzia sandstone (Möller 1987), are not of importance to the general question of trilobite or non-trilobite origin for the trace. Following the discussion of Bergström (1973a), Hesselbo (1988) and Jensen (1990), an olenellid trilobite is a likely producer. The above-described impressions of genal spines further support this interpretation. The scratch pattern indicates a leg with two strong distal claws and a set of finer setae (or claws) on the anterior side (Bergström 1973a, pp. 53–54). However, the leg morphology is unknown in olenellid trilobites. *Rusophycus* is not by definition a trilobite burrow (Bromley 1990). Bromley & Asgaard (1979) presented several Triassic freshwater burrows, probably made by notostracan branchiopods, with a range in morphologies covering several species of early Palaeozoic supposed trilobite burrows. One of these (Bromley & Asgaard 1979, Fig. 17b), is a miniature replica of *R. dispar*. However, I concur with Bromley & Asgaard (1979) in placing very small *Rusophycus* in a separate species (see *R. eutendorfensis*). The ethological motivation of *Rusophycus* burrows has been extensively discussed (e.g., Seilacher 1953, 1985;

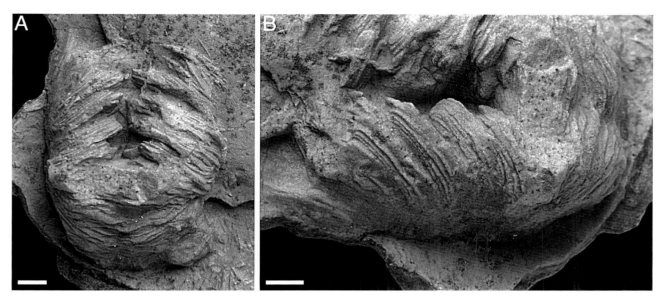

Fig. 57. □A. *Rusophycus dispar* with widely separated proverse scratch marks. Scale bar 10 mm. Lugnås. Specimen in collection of Holger Buentke, Lugnås. □B. Detail of proverse scratch marks in A. Scale bar 10 mm.

Osgood 1970). Schmalfuss (1981) and Seilacher (1985) depict a feeding chamber, formed below the carapace of the burrowing trilobite, in which food particles were extracted by the outer ramus from sediment brought into suspension. In the Mickwitzia specimens, evidence for hunting of soft-bodied animals has been found (Bergström 1973a, Jensen 1990) and is considered a likely, if not the sole, motivation for most *R. dispar*.

Rusophycus jenningsi (Fenton & Fenton, 1937)

Fig. 58A–C

Material. – Figured slab, RM X3347. Interval B, 30–60 cm above basement, at Hjälmsäter. Additional specimens in collection of Jan Johansson, Sköllersta.

Description. – Deep burrow with a U-shaped vertical profile (Fig. 58C). Sides of burrow steep, originally probably nearly vertical, but compactionally skewed (Fig. 58B). In its deepest part, a shallow median furrow divides the trace into two lobes. These lobes have poorly preserved, coarse scratch marks, meeting at about 180° (Fig. 58A–B). To one end of the burrow, herein called posterior, these scratches give way to finer scratches with a lower V-angle. The latter reaches about halfway to the top of the burrow. At the opposite end the scratch marks are bordered by a faint crescentic rim. Laterally bordering the coarse scratches and continuing up the sides of the trace in a near vertical orientation are two approximately parallel ribs,

with a relief of up to 0.5 mm (Fig. 58B). Similar but smaller U- to J-shaped ribs, including some with diverging orientation, are found higher up on the trace (Fig. 58C). On the sides of the burrow are densely set finer ridges with a mainly vertical direction (Fig. 58B). These occur in sets of at least five, with set widths of about 3 mm. Indifferent preservation prevents observation of their relative abundance on different sides of the trace. Depth of trace 5.5 cm, length at top of burrow 6.5 cm, width about 4 cm.

Discussion. – This burrow compares well with *Rusophycus jenningsi* (Fenton & Fenton, 1937) from the Lower Cambrian of Alberta, which is a deep burrow with vertical ribs, cephalic impressions and two(?) types of scratches. Seilacher (1970) and Bergström (1973a) interpreted the transverse coarse scratches in the deepest part of *R. jenningsi* as made by claws of the inner ramus of a trilobite and the posterior, longitudinal and finer scratches as made by the outer ramus. Later, Bergström (1976, p. 1627) modified the interpretation by stating that no definite inner-ramus scratches are seen in *R. jenningsi*. The scratch pattern in the Mickwitzia sandstone specimen also presents some problem. The deep, coarse scratches are similar to those seen in Mickwitzia sandstone specimens of *Cruziana rusoformis* and to the transverse scratches in *R. dispar*. The finer scratches have some similarity to markings occasionally seen in proverse parts of *R. dispar* and in Arthropod scratch mark Type A. Bergström (1973a, p. 53) identified series of up to five fine parallel scratches in front of the main claws in a well-pre-

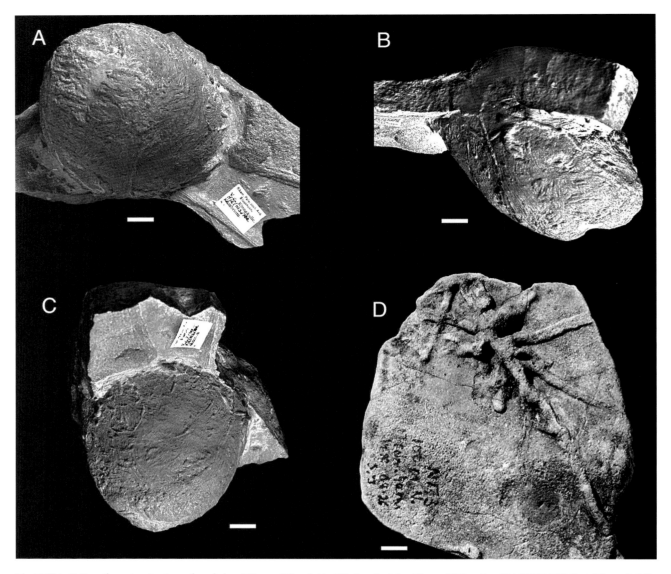

Fig. 58. □A–C. *Rusophycus jenningsi* seen from below (A), rear (B) and side (C) view. Scale bar 10 mm. Hjälmsäter. RM X3347. □D. *Scotolithus mirabilis.* Specimen preserved on top of sandstone slab. Several burrows radiate from an inferred vertical central stem. Scale bar 10 mm. Trolmen. RM X3334.

served specimen of *R. dispar*. This lead him to postulate a number of setae on the anterior side of the inner ramus. Though the two types of scratches in this trace could possibly have been caused by inner and outer ramus, they may also be the work of the inner ramus only. Compared to *R. jenningsi*, as illustrated by Fenton & Fenton (1937, Fig. 1), the Mickwitzia sandstone specimen has less pronounced cephalic-shield impressions. Whether the head shield took an active part in the excavation of this burrow is uncertain, but the position of the shield impression shows a head-down posture of digging. The diverging orientation of some lateral ribs reveals that at least during the

early stages of digging there was some horizontal shift of body orientation.

A similar burrow was named cf. *Cruziana jenningsi* by Seilacher (1970, Fig. 6A) but later identified as *Rusophycus leifeirikssoni* Bergström, 1976 (Seilacher 1985, Fig. 1E). According to Bergström (1976, p. 1626), *R. leifeirikssoni* was made solely by the inner ramus and is characterized by narrowing in its deepest part, taken to indicate a tail-down direction of the trilobite. The specimen described here shows no marked narrowing in its deepest part, and the crescentic marking in the base of the burrow was most likely caused by the rim of a head shield.

Rusophycus eutendorfensis (Linck, 1942)

Figs. 31, 32B, 67C

Synonymy. – See Bromley & Asgaard (1979, p. 64).

Material. – Specimens from Billingen, Lugnås and Hjälmsäter. Interval B and D. Figured slabs, SGU 8586, RM X3303, RM X3344. Additional specimens in collection at RM

Description. – Small bilobate burrows, buckle- or heart-shaped, with steep lateral borders (Fig. 32B). Lobes mostly smooth and separated by a distinct median furrow. Fine transverse or gently inclined striation with ridges fractions of a millimetre (in one specimen measured as 0.1 mm) apart are occasionally seen. Dimensions of a prominent specimen: length 4 mm, width 2.4 mm, depth 1.8 mm, but most specimens are smaller. May be arranged in loose, straight to somewhat winding chains of burrows, and pass into *Cruziana tenella*.

Discussion. – This type of burrow has been widely reported as *Rusophycus didymus* (Salter, 1856). *Arenicola didyma* Salter, 1856, was erected for paired elongated hollows from the Longmyndian of Wales. Crimes (in Crimes *et al.* 1977, p. 108) accepted an organic origin for the type material of *Arenicola didyma* but questioned its affinity to younger, unquestionably biogenic specimens assigned to *Rusophycus didymus*. Examination of specimens in the collections of the Sedgwick Museum, Cambridge, indicates that they are different from the supposed arthropod burrows; the slits are strongly parallele, often not paired and may be arranged along a wavering path. According to Bland (1984), these are impressions belonging to the problematicum *Arumberia*. It seems best to avoid Salter's species for this type of burrow. I follow Bromley & Asgaard (1979) in including specimens assigned to *R. didymus* in *R. eutendorfensis* (Linck, 1942). Bromley & Asgaard (1979) used *R. eutendorfensis* for *Rusophycus* of variable morphology, with the prime characteristic being their small size: width given as 0.8–14 mm. In the Mickwitzia sandstone *R. eutendorfensis* is closely linked to *Cruziana tenella*.

Ichnogenus *Scotolithus* Linnarsson, 1871

Diagnosis. – Curved burrows running down from a vertical shaft, at the base of bed, typically turning horizontal. Arranged in a radial or fan-shaped form. Number of burrows low, not densely covering the sediment surface, and due to burrow curvature resulting in an untouched area immediately below the vertical shaft. (Partly based on Linnarsson 1871.)

Type species (by monotypy). – *Scotolithus mirabilis* Linnarsson 1871

Scotolithus mirabilis Linnarsson 1871

Figs. 36A–B, 58D, 59

Synonymy. – □1871 *Scotolithus mirabilis* n.g. et sp. – Linnarsson, pp. 18–19, Fig. 21, *non* 22.

Diagnosis. – As for ichnogenus.

Type material. – Of the type material of *Scotolithus mirabilis* Linnarsson, 1871, only the specimen (RM X179) that Martinsson (1965) selected to be a lectotype has been found. As discussed below, this specimen does not show the typical morphology of *Scotolithus mirabilis*.

Well-preserved material, very similar to that illustrated by Linnarsson (1871), is found in the roof of abandoned mines at Älerud, Lugnås. This material is not collectable. Conforming to the description of Linnarsson (1871) are also specimens on a fractured block from Trolmen harbor (RM X3346a, b) (Fig. 36A–B). This slab has a poorly preserved plan view but on the vertical fracture planes shows the characteristic vertical shaft from which emanate curved, radiating burrows.

Other material. – RM X3334. Several specimens in situ, 3.6 m above basement in the roof of abandoned mines at Älerud (including 'Minnesfjället', see Fig. 2A), Lugnås.

Description. – Radiating burrows diverging from a vertical, common shaft, gradually turning horizontal, typically resulting in untouched central area below the vertical shaft (Fig. 59). Burrows are essentially cylindrical or with

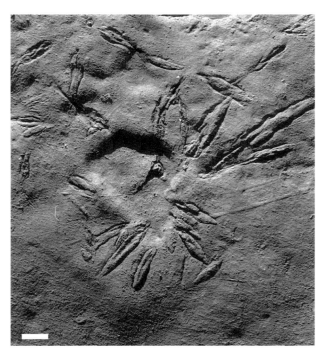

Fig. 59. Scotolithus mirabilis on sandstone sole from quarry at Älerud, Lugnås. Level B. Scale bar 20 mm.

a spreite of a few laminae. Burrow walls have varying surface sculpture, reflecting consistency of sediment. Preserved as full-relief structures in sandstone, sand-filled structures in clayey–silty sediments and as negative epireliefs (Figs. 36A–B, 58D, 59). Terminations of the burrows typically pointed, as a consequence of inclination to bedding. One slab (Fig. 36A–B) has two specimens preserved as elongated cylindrical to conical burrows spreading out in shape of a loose broom; at the center the burrows are directed vertically; away from center they are inclined, and some turn horizontal at base of slab. Walls with distinct dark lining. Most burrows are preserved as hollows; where fill is present it differs from surrounding sediment. The smaller of the two specimens shows two conical burrows along fracture, while the larger specimen has at least eight burrows. Diameter of burrows 2–5 mm. The whole burrow system may span several decimetres.

Taxonomy. – Martinsson (1965, p. 219) erected a lectotype for *Scotolithus mirabilis* from one of the specimens illustrated by Linnarsson (1871), and considered it synonymous with *Halopoa imbricata* and *H. composita*. This was followed by Häntzschel (1975). Linnarsson (1871) described *Scotolithus mirabilis* in terms of its burrow ornamentation as well as its overall burrow morphology, though only one of the illustrated specimens shows the former. The burrow morphology was described as radially arranged branches, the meeting point of which is not visible. Within the rock, these branches connect to a vertical stem demarcated from the surrounding rock by a certain difference in color and consistency (Linnarsson 1871, p. 19). Linnarsson (1871, p. 19) considered the specimen subsequently chosen by Martinsson (1965) as lectotype to be a possible branch of *Scotolithus mirabilis* embedded in a differing attitude. Of the two characters, the wall ornamentation and overall burrow geometry, the latter is of higher ichnotaxonomic importance. The burrow-wall ornamentation is similar to that in *Halopoa* (*Palaeophycus*), but, as discussed in connection with that genus, the burrow sculpture reflects sediment consistency. Thus, of the two specimens illustrated by Linnarsson (1871), only one shows the more important characters typical of the species, and the lectotype chosen by Martinsson (1965) is therefore not representative.

The specimen illustrated by Linnarsson (1871, Fig. 21) shows about eight burrows radiating out in an area covering about half a circle. Other specimens have fewer burrows covering a smaller angle. As noticed by Linnarsson, the burrows pinch out rapidly, indicating a rapid shift in orientation towards the vertical. These burrows may be compared with species of *Phycodes* and also with rosetted burrows of *Asterichnus* Bandel, 1967, type. These burrows differ from most species of *Phycodes* by having a vertical main trunk. The main exception is *Phycodes flabellum* (Miller & Dyer, 1878) from the Ordovician of Cincinnati,

which has burrows typically arranged in a flabellate pattern, originating from an L-shaped structure with the long side in vertical orientation. The burrows are tightly grouped and show annulations. However, a greater similarity exists with *Asterichnus lawrencensis* Bandel, 1967. In the type area (Upper Pennsylvanian, Kansas) it is preserved as radiating burrows on tops of rippled sandstones (Bandel 1967). A central knob was interpreted to be part of a vertical tube that led up into the overlying sediment (Bandel 1967, p. 3). Bandel (1967) noted that *Asterichnus* is arranged in linear patterns and that connecting trails were found between a few specimens. A vertical burrow was only inferred from the type material of *Asterichnus* and to my knowledge this is the case also for other reports (Eagar *et al.* 1985; Chamberlain 1971; Seilacher 1983b; Biron & Dutoit 1981). It is not clear whether the similarities between *Asterichnus* and *Scotolithus* are superficial or not. *Scotolithus* differs from *Dactyloidites* in the absence of a well-developed spreite.

Interpretation. – This trace was, at least in some cases, made by the same animal(s) that made *Palaeophycus imbricatus*, since *Scotolithus mirabilis* occasionally forms radiating patterns along the horizontal burrows of *P. imbricatus* (Fig. 58D). This can be explained as a more intensive search in certain areas of the sediment, whereby a vertical shaft was constructed, probably to bring contact to oxygenated water.

Ichnogenus *Skolithos* Haldeman, 1840

Skolithos linearis Haldemann, 1840

Fig. 5A

Material. – Common in outcrops of interval E at Hällekis. Figured slab RM X3289 from Hällekis. Also observed in Mossen core.

Description. – Vertical tubes, up to several decimetres long, of about constant width except for occasional funnel-shaped top, the depth of which is small compared to length of tube. The sand within the tubes is generally similar to that in the surrounding sediment and shows no evidence of biogenic influence. Sediment in tube demarcated by clayey zone typically with oxidized iron minerals. Diameter of tube about 5 mm. Length up to several decimetres. Funnel-shaped top may expand to a width of 20 mm.

Discussion. – These burrows are assigned to *Skolithos*, though some have a small terminal funnel which is a character of *Monocraterion* (e.g., Alpert 1974b). Durand (1985a, b) admitted a small terminal funnel in *Skolithos*, a procedure followed here. These burrows occur in amalgamated sediments, and it is acknowledged that *Skolithos* and *Monocraterion* may be the same burrow made under differing sedimentological conditions (see discussion on *Monocraterion*). The specimens here assigned to *Skolithos* are vertical tubes with a width that is small compared to the length. This and the vertical orientation suggest assignment to *S. linearis* (Alpert 1974b).

Ichnogenus *Syringomorpha* Nathorst, 1886

Syringomorpha nilssoni Torell, 1868

Fig. 60

Material. – Three slabs collected with about 20 specimens; figured RM X3335, with three well-preserved specimens. Additional material observed on loose material from Hällekis and in the collection of Jan Johansson, Sköllersta. Observed in outcrop, 5.9 m above basement in a quarry at Älerud, Lugnås.

Description. – Closely stacked burrows about 2 mm wide, with a nearly vertical orientation; mainly straight or gently curved, at their base turning horizontal at a common level, thereby truncating preceding burrows (Fig. 60).

Observed horizontal extension of spreiten less than 10 cm, with a straight or slightly curved course. In horizontal sections, individual burrows are visible as crescent-shaped lenses, separated from each other and the surrounding sediment by clayey, dark material (Fig. 60A). Fill identical to surrounding sediment.

Discussion. – Westergård (1943, p. 28) reported *Syringomorpha nilssoni* from a block at Råbäck harbor, suspected to belong to the Mickwitzia sandstone. Loose blocks with *Syringomorpha* are also found at Trolmen. The specimen observed *in situ* from Lugnås is found in interval C, and the lithology of blocks from Hällekis points to this interval also for the occurrence at Kinnekulle.

The orientation of *Syringomorpha* relative to bedding has received some confusing notions. Nathorst (1874, p. 45) thought *Syringomorpha* to be made by whipping movements of a plant, implying it to be parallel to bedding. On the other hand, Holst (1893, p. 10) reported specimens from the Kalmarsund area to be consistently vertical. It is therefore possibly a semantic misunderstanding that lead Richter (1927, p. 262) to cite Holst (1893) in support of *Syringomorpha* having an orientation independent of layering. A vertical orientation of the spreite is consistent for all examined occurrences of *Syringomorpha* in Sweden. However, Richter's (1927) notion has been influential (e.g., Häntzschel 1975) and has lead to identification of horizontal forms as *Syringomorpha*

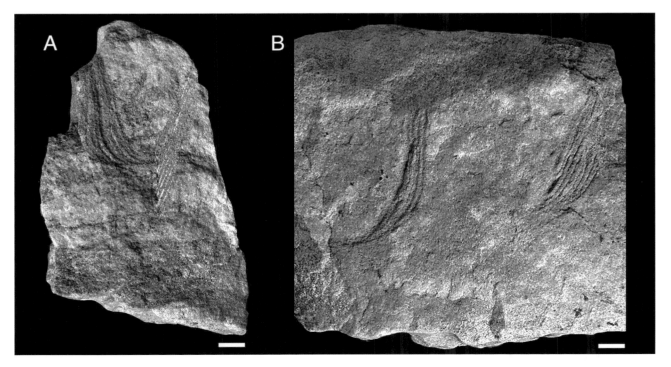

Fig. 60. Syringomorpha nilssoni from the Mickwitzia sandstone at Hällekis. Probably interval C. Scale bar 10 mm. RM X3335. □A. Fractured surface showing specimens with burrows curving to a horizontal orientation at the base. Hällekis. □B. Two specimens seen in side view. Hällekis.

Fig. 61. Teichichnus ovillus □A. Field photograph of large specimen orientated upside-down. Hällekis. Probably interval D. □B. Specimen with a relatively closely stacked spreite. Oriented upside-down. Scale bar 10 mm. Hjälmsäter. RM X715. □C. Hällekis. Scale bar 10 mm. Interval D. RM X3338.

(Gibson 1989). The Mickwitzia sandstone specimens agree well with material illustrated by Torell (1868, Pl. 2:2), which is unusual in showing two specimens with burrows inclined nearly at right angle to each other. This material appears to be lost, and a neotype was chosen from glacial erratic material by Pek & Gába (1983), but not described or illustrated.

The burrows within the spreite of *Syringomorpha nilssoni* were added with little shift of vertical level. This is indicative of a horizontally progressing dwelling or feeding–dwelling burrow (cf. Durand 1985a).

Ichnogenus *Teichichnus* Seilacher, 1955

Teichichnus ovillus Legg, 1985

Fig. 61

Synonymy. – □1965 Large teichichnian burrows – Martinsson, Figs. 21–23. □1985 *Teichichnus ovillus* n. ichnosp. – Legg, p. 161. Pl. 4A.

Material. – Six slabs with 12 specimens; figured slabs RM X715, 3338. Additional specimens observed in the field, especially common in interval D.

Description. – Straight to gently curved stacked gutter-shaped laminae forming low wall-like structures (Fig. 61). In collected specimens the number of laminae is low, with a dominating retrusive arrangement, though protrusive development also occurs. One specimen observed in the field has a height of about 5 cm and a length exceeding 50 cm (Fig. 61A). In vertical plan the longitudinal outline of the laminae is straight or curved, usually with asymmetric development with one more strongly curved and one more horizontal end. Burrows occur as isolated lenses of fine sandstone with sparse grains of coarser material, inside clayey beds; occasionally they pass through sandy beds. The surface may have an ornamentation of longitudinally directed ridges. In the specimen where these are best preserved, the ridges are spaced at about 2 mm.

Discussion. – These burrows bear strong similarity to large burrows described by Martinsson (1965) from the Middle Cambrian of Öland. Martinsson (1965) used the term teichichnians to include a wide range of burrows (Martinsson 1965), some of which are not referable to *Teichichnus*. The type species of *Teichichnus*, *T. rectus* Seilacher, 1955, has a retrusive, tightly guided spreite (Seilacher 1955b). Legg (1985) erected *Teichichnus ovillus* for burrows with both protrusive and retrusive spreite that are less straight-walled than *T. rectus*. Most Mickwitzia sandstone specimens are assignable to *T. ovillus* in having a spreite that is low in relation to its length, not closely stacked, and with a longitudinal asymmetry. Ichnospecies of *Teichichnus* and related ichnogenera are in need of a monographic treatment in order to establish the relative importance to be put on spreite proportions, manner of stacking and surface ornamentation. See also discussion on *Trichophycus*.

The basic construction was flat U- or L-shaped burrows that were shifted upwards and in the direction of burrow length, as a consequence of sediment influx.

Ichnogenus *Treptichnus* Miller, 1889

Treptichnus bifurcus Miller, 1889
Fig. 62A

Material. – Two slabs collected; figured RM X3337. Interval B, Lugnås.

Description. – Trace fossil consisting of a series of more or less straight segments joined to each other at an angle, generally intersecting some distance from end of preceding segment (Fig. 62A), to form projections. These projections curve upwards; typically they are somewhat swollen compared to the main burrow. Where the overall course of the burrow is straight, segments typically alternate in direction, forming a zigzag pattern that may lack projections. Where the burrow curves, successive segments are placed on the inside of the preceding ones, in the direction of curvature. Course of burrow system winding, also with crossings (Fig. 62A); length of segments 5–15 mm; diameter of segments 1–3 mm; length of projections 3–6 mm. Except for short segments preserved in hyporelief, the trace is preserved as full reliefs pressed onto base of sandstone (Fig. 62A). Fill identical to that of slab.

Remark. – Archer & Maples (1984, Fig. 6) and Maples & Archer (1987, Fig. 5) showed that lower-plane preservation of *Treptichnus bifurcus* results in a zigzag trail lacking projections. This preservational variation, known as *Plangtichnus erraticus* Miller, 1889, Maples & Archer (1987) maintained as a separate ichnogenus. The same preservation is found also in the Mickwitzia sandstone material but is here not assigned a separate name (see also Buatois & Mángano 1993).

Treptichnus pedum (Seilacher, 1955)
Fig. 62B

Synonymy. – □1955 *Phycodes pedum* n.sp. – Seilacher, pp. 386–388. Figs. 4A, B; 5:13; Pls. 23:7; ?23:6; Pl. 25:3.

Material. – Two slabs in collection of SGU; figured SGU 8598. Probably interval B. Lugnås.

Description. – Trace consisting of rows of smooth, curved burrows, joining each other at low angles and intersecting to form projections (Fig. 62B). Path highly winding, including looped sections. Low-angle, single-sided addition occurs also in straight sections; in places the segments are arranged in a nearly straight succession. A zigzag pattern is only rarely developed. Length of segments 8–13 mm, of which 1–4 mm constitutes projections. Preserved as hyporeliefs with fill identical to that of slab.

Treptichnus and Phycodes pedum. – The type ichnospecies *Treptichnus bifurcus* Miller, 1889, originates from Carboniferous (Namurian) laminated siltstones of the Hindostan Whetstone Beds of Indiana (Miller 1889). The figured specimen (Miller 1889; Fig. 1095) consists of an irregularly running zigzag trail with short projections at each angle. Where the course of the burrow is straight, segments alternate in direction; where the course is curved, the projections are directed outward. Additional material from the type area has recently been presented by Archer & Maples (1984) and Maples & Archer (1987). This shows the trace to be a three-dimensional burrow system, with different morphologies resulting from the horizontal level at which the trace cuts across bedding planes (Maples & Archer 1987, Fig. 5). Despite Miller's (1889) description and illustration, *Treptichnus* has been used virtually only for traces with a straight course and with segments of regularly alternating direction (e.g., Häntzschel 1975). As suggested by Maples & Archer (1987), this probably depends on *Treptichnus* being identified with the 'feather-stitch trails' of Wilson (1948), which are more or less straight zig-zag traces. Maples & Archer (1987, p. 894) pointed out that this trace is clearly different from the type material of *Treptichnus*, most notably in being preserved in two planes. It appears to be a horizontally spiraled trace. Several authors have observed that *Phycodes pedum* Seilacher, 1955, may have portions of *Treptichnus*-type (e.g., Fritz & Crimes 1985; Bryant & Pickerill 1990). Burrows identical to that described here have been assigned to *Phycodes pedum* by Crimes *et al.* (1977), Fritz & Crimes (1985), Bryant & Pickerill (1990), Walter *et al.* (1989), and others.

Osgood (1970, p. 342) noted that *Phycodes pedum* differs from other species of *Phycodes*, such as the type species *Phycodes circinatum* Richter, 1853, and that it merits a new ichnogeneric designation. Most reports of *P. pedum* are of specimens with a behavior pattern that is similar to that in *Treptichnus*, with only minor morphologic differences, such as poorly developed alteration in segment orientation. *P. pedum* as used today includes a wide range of forms. Where there is a long chain of burrows (e.g., Bryant & Pickerill 1990), assignment to *Treptichnus* is logical. This is less clearly so in short specimens (e.g., Narbonne

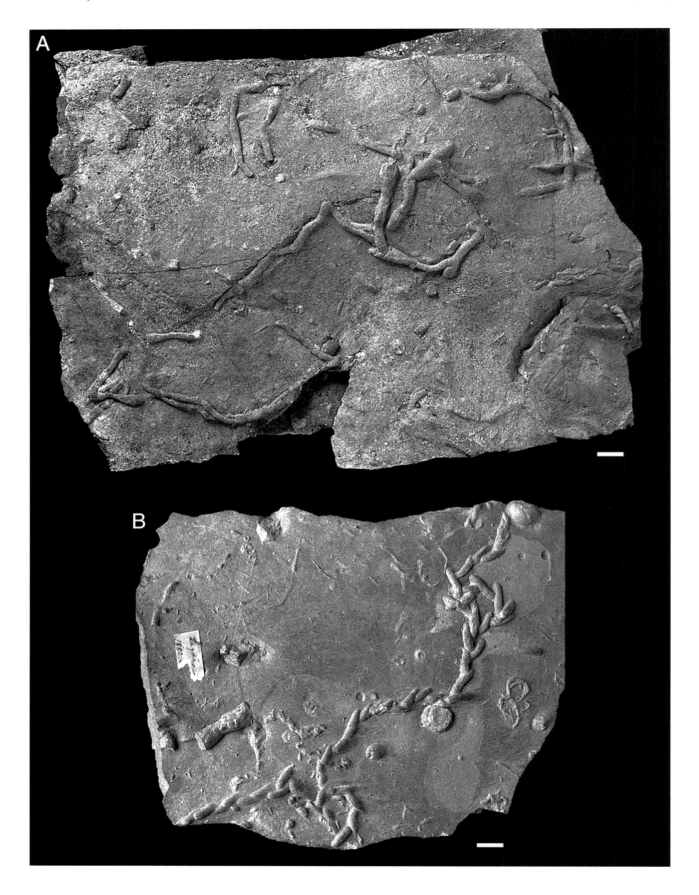

et al. 1987. Fig. 6E). Seilacher (1955b, Fig. 4b) reconstructed *P. pedum* as a flat U-tube with one fixed and one gradually expanding end with successive up-turned burrows arranged on the outer side of a curving path. He did not state whether the expanding U-tube was inferred or observed, but on examining the illustrations (Seilacher 1955b, Pl. 25:3) it seems equally likely that *P. pedum* was formed by addition of segments, in a treptichnian manner. A weakly developed alternation in direction is actually seen also in the holotype of *P. pedum* (see Seilacher 1955b, Pl. 25:3), and *P. pedum* is therefore assigned to *Treptichnus*.

Another species doubtfully referred to *Phycodes*, is *Phycodes? antecedens* Webby, 1970, which shows similarity to *P. pedum* Webby, 1970. *Yangziichnus* Yang, 1984, a Lower Silurian form from Yangzi, China (Yang 1984), and the similar *Phycodes templus* Han & Pickerill, 1994, from the Lower Devonian of New Brunswick (Han & Pickerill 1994) may also warrant comparison with *Treptichnus*.

At present, there are four ichnospecies of *Treptichnus*: *T. bifurcus* Miller, 1889, *T. triplex*, Palij, 1976, *T. lublinensis* Pacześna 1986, and *T. pollardi* Buatois & Mangano, 1993. *T. lublinensis* differs from the type species of *Treptichnus*, in having densely arranged segments, set at a high angle to each other (see Pacześna 1985, 1986). *T. triplex* has segments with three longitudinal ridges (see Palij *et al.* 1983; Fedonkin 1985). *T. pollardi* lacks projecting segments and has vertical shafts expressed as pits at bedding planes. As proposed by Jensen & Grant (1992), *Phycodes pedum* is tentatively considered a fourth ichnospecies of *Treptichnus*, having relatively shorter, more closely set, curved segments, and a less ordered arrangement of segments. There may be single-sided, often very dense branching, also where the course is straight, which is only rarely seen in *T. bifurcus* (see Maples & Archer 1987, Fig. 2:4).

Geyer & Uchman (1995) did not accept this, since they regarded *Treptichnus* to be more regular and symmetric and to be permanently open burrow-systems lacking sediment reworking. The original material of *Treptichnus* seems to answer the first point. I find that deciding whether a burrow was permanently open or back-filled is in most cases not possible. Trace fossils, by their very nature, do not lend themselves to pegging into distinct morphological categories, and this may be particularly true in the Cambrian, as exemplified in this study. I maintain, however, that most *Phycodes pedum* can be assigned to *Treptichnus*.

Treptichnus is generally interpreted as a systematic feeding structure with each segment reaching up to the sediment surface (Seilacher & Hemleben 1966). The Mickwitzia sandstone specimens have a passive fill and show signs of being exhumed by the casting sediment. The relative thickening of the projections in *T. bifurcus* is probably caused by flattening of vertically directed parts during compaction. Though there may have been backfill that was washed out in the process of casting, this trace can also be interpreted as a series of dwelling tubes inhabited by a detritus feeder (sensu Bromley 1990, pp. 6–7) feeding on detritus surrounding the burrow opening.

Ichnogenus *Trichophycus* Miller & Dyer, 1878

Trichophycus venosus Miller, 1879

Fig. 63

Synonymy. – □1990 Worm-burrows – Jensen, pp. 31–32, Figs. 2–4, 6A–B, ?6C.

Material. – Figured slabs, LO 6698t, RM X3336. Additional specimens observed in Level D at Hällekis and Hjälmsäter.

Description. – Flat, longitudinally U- to L-shaped, straight to slightly curved burrows, often possessing an incipient, mainly retrusive spreite (Fig. 63A). The spreite generally consists of a few, not very closely packed, gutter-shaped laminae, and there may be short segments that deviate laterally from the main course of the burrow. The lower surface of the burrow may have longitudinal parallel striae less than 1 mm apart (Fig. 63A–B). Occasionally, there are also transverse or oblique, generally finer striae. A few specimens have rounded, button-like endings, possessing radiating patterns of more or less straight striae. The traces occur as sand-filled structures in clayey surrounding, with sediment similar to that in the overlying bed. Contact with sole of sandstone is often through abruptly upturned ends. The traces may occur closely grouped, also showing cross-cutting relations. Dimensions of a well-preserved specimen: length 15 cm; width 1.5 cm; depth 4 cm. Observed variation in width 1.4–1.7 cm.

Discussion. – These burrows have features which ally them to *Teichichnus* as well as *Trichophycus*. As noted by Frey & Chowns (1972), Bergström (1976), and Frey & Howard (1985), differentiation between these burrows may be problematic. Seilacher (1955b, p. 379) erected *Teichichnus* to include 'alle mauerförmig gestreckten Spreitenbauten'. A spreite is typically present also in *Tri-*

Fig. 62. □A. *Treptichnus bifurcus*, with typical development best seen in looped part. Straight parts with an undulating surface are similar to 'Hormosiroidea' isp. Scale bar 10 mm. Lugnås. RM X3337. □B. *Treptichnus pedum*, specimen with densely set branches in places with zigzag development. Also seen are small contorted burrows Scale bar 10 mm. Lugnås. SGU 8598.

Fig. 63. Trichophycus venosus. □A. Several flatly U-shaped burrows, cylindrical or with a low spreite. Longitudinal ornamentation faintly visible. It is uncertain if the specimen in the lower right corner, with a flat median furrow, is a *Trichophycus* Scale bar 10 mm. Locality unknown. LO 6698t. □B. Lower surface of specimen with low gutter-shaped spreite having longitudinal ridges. Scale bar 10 mm. Interval D, Hjälmsäter. RM X3336.

chophycus, which has been characterized by fine scratch-mark pattern on the lower surface (e.g., Seilacher & Meischner 1964; Häntzschel 1975; Seilacher 1983b). The visibility of this structure is governed by the state of preservation and, as scratch marks also have been reported in *Teichichnus*, it is far from an ideal criterion for differentiation. It may be more feasible to use the overall shape of the burrow and spreite. *Trichophycus* may have vertical branching (Osgood 1970; Frey & Chowns 1972), and the spreite in *Teichichnus* is 'more tightly knit and more nearly planar in replication' (Frey & Howard 1985, p. 391) than in *Trichophycus*. Where a spreite is absent or weakly developed, *Trichophycus* grades into *Palaeophycus* (Frey &

Chowns 1972). Transition into *Phycodes* is also possible, especially *Phycodes palmatus* (Hall, 1852), which typically has a more pronounced lateral branching. Seilacher & Meischner (1964) and Osgood (1970) suggested that the spreite of *Trichophycus* resulted from material being excavated from the roof of the burrow and packed onto its floor. In the Mickwitzia sandstone specimens, the filling material is similar to that in the overlying sand or siltstone, and therefore Jensen (1990) suggested that the material drifted in during the animal's mining activity. However, it is equally possible that the spreite was made in response to sediment influx and that it is a dwelling or dwelling–feeding structure. As such, the morphology of

Trichophycus would largely depend on the rate of sediment influx and the animal's capability to react to this influx; hence it could be expected to show high variability.

Here, *Trichophycus* is used in a slightly modified sense to that of Frey & Chowns (1972), who applied it for broad, irregular U- or L-shaped burrows, unbranched or with vertical or minor lateral branching, some having spreiten; walls distinct, typically ornamented with scratch marks. Several specimens of flat U- to L-shaped spreiten burrows from the Mickwitzia sandstone were illustrated by Jensen (1990), but their ichnogeneric assignment was left undecided because of lack of details on their ventral surfaces. Following the discussion above, most of these specimens may be referred to *Trichophycus*. The morphology of these traces is dependent on the burrow material; specimens from Hällekis, where the fill is finer-grained, are more flattened owing to compaction and have less well-preserved scratches than specimens from Lugnås. Possibly assignable to *Trichophycus* are some of the Middle Cambrian burrows from Öland given as teichichnia by Martinsson (1965, Figs. 21–24). These are very large structures with retrusive as well as protrusive spreiten (Martinsson 1965). Similar but smaller specimens are found in the Mickwitzia sandstone (see *Teichichnus ovillus*) and also in the Norretorp sandstone near Brantevik, Scania. *Trichophycus* grades into *Teichichnus ovillus*, and several specimens are intermediate between the two.

In the Cincinnati area of Ohio, Osgood (1970) recognized two species; *Trichophycus venosum* Miller, 1879, and *T. lanosum* Miller & Dyer, 1878. *T. lanosum* was said to differ by having a small depression with radiating striae and by its sinuous nature. However, judging from Fig. 68:8 of Osgood (1970), the implied sinuosity is highly doubtful and could well result from a number of close parallel burrows. As *Trichophycus lanosum* is known from only two specimens, it seems possible that *Trichophycus lanosum* and *Trichophycus venosum* are synonymous. *Trichophycus thuringicum* Volk, 1968, from the Ordovician of Thuringia, Germany, is a cylindric burrow with coarse longitudinal scratches (Volk 1968). Most reports of *Trichophycus* have been from Ordovician strata, including Wales (Pickerill 1977), Norway and Iraq (Seilacher & Meischner 1965), Germany (Volk 1968), Georgia, USA (Frey & Chowns 1972), Ohio, USA (Osgood 1970), and Newfoundland, Canada (Seilacher & Crimes 1969; Bergström 1976). Seilacher (1983b) extended the range to the Carboniferous on the basis of finds from Egypt. These reports have generally been at the genus level only. The Mickwitzia sandstone specimens fall within the range of *Trichophycus venosus*. Where scratch pattern is visible it is similar to that of specimens from Cincinnati (Osgood 1970) and Newfoundland (Fillion & Pickerill 1990a).

Arthropods are usually considered to be the producer of this type of trace, including trilobites, or annelids (Frey & Chowns 1972; Seilacher 1983b). Some of the Mickwitzia sandstone specimens have a scratch pattern indicating a spinous, more or less radially symmetric digging apparatus, making priapulid worms a possible candidate (Jensen 1990).

Ichnogenus *Zoophycos* Massalongo, 1855

Zoophycos (*Rhizocorallium?*) isp.
Fig. 64

Material. – Figured slabs RM X3225, 3228, with about 18 specimens. Hjälmsäter, 0.4–1.0 m above the basal gneiss. One slab in collection of Jan Johansson, Sköllersta.

Description. – Circular to heart-shaped flat spreite consisting of arcuate to S-shaped ridges radiating from about the center of the trace (Fig. 64). The spreite is enclosed by a semi-cylindrical, rarely cylindrical, marginal tube which runs from the center out to the limit of the spreite, where it sharply turns and follow the outline of the spreite. The spreite is of very low relief, with the ridges delineating shallow troughs; relief of the marginal burrow is consistently higher. Maximum width of spreite 20–61 mm; width of marginal tube about 2 mm. Diameter of spreite and marginal burrow show positive correlation (based on nine best preserved specimens). Preserved on lower surface of sandstone. Sections reveal no vertical continuation of the trace into sandstone.

Discussion. – The lack of vertical extent and a general washed-out appearance of the marginal tube indicate that these traces were preserved through excavation and casting of a trace made in clayey sediment. This may explain the difference between the spreite and marginal tube; the former is visible only as imprints of filled spreite, while the marginal tube was an open structure, or consisted of easily exhumed material. The mode of preservation makes three-dimensional reconstruction difficult. An enclosing marginal tube is indicative of a basal U-shaped burrow, which may have been helicoidal or flat (cf. Wetzel & Werner 1981). The wide range in spreite diameters of crowded *Zoophycos* (*Rhizocorallium?*) isp., could depend on the casting level cutting a conical helicoidal spreite at different vertical levels, and the size correlation of spreite and marginal tube may be a growth vector. Usually one side of the trace is better preserved, which also argues for a helicoidal construction. Some authors prefer to place small circular forms with concave-up spreite in *Spirophyton*, in the sense of *Spirophyton eifliense* Kayser, 1872 (Simpson 1970; Miller & Johnson 1981). Though meeting the first two criteria, the Mickwitzia sandstone specimens appear to have a rather flat spreite. As they are smaller than typical *Zoophycos*, an affinity to *Spirophyton* may still be possible. Kotakes' (1989) find of specimens that in the

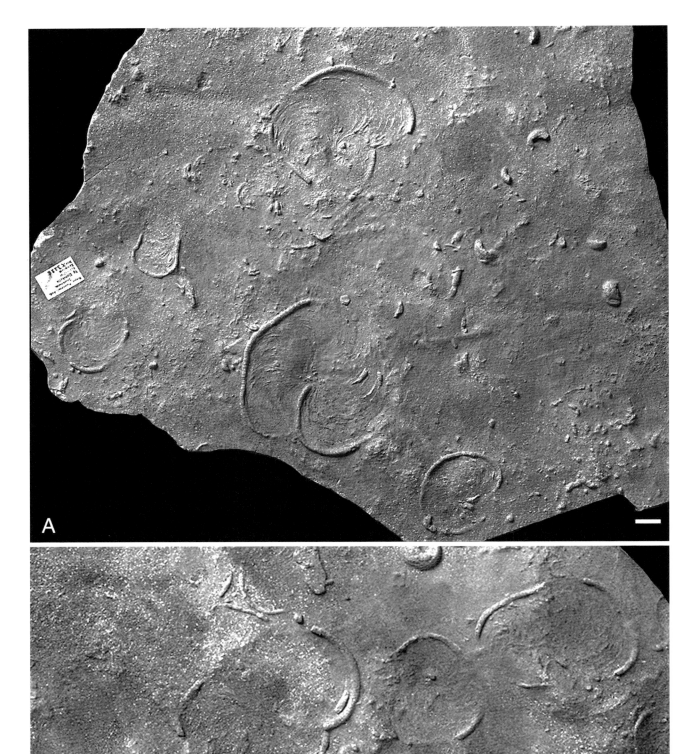

upper part are *Spirophyton* and in their lower parts gradually turn into *Zoophycos* may question the ichnotaxonomic significance of an upturned spreite.

This form may be a planispiral (?and trochospiral) form of *Rhizocorallium*, rotating around one limb (Fig. 22C). The width of the spreite would thus correspond to the width of the basic U-shaped burrow. This interpretation explains the morphology of the spreite, with the radial tube corresponding to the final position of the burrow. If it is assumed that the modified U-tube was maintained open, as indicated by its preservation, it's function to provide oxygen-rich water may be questioned. However, the same problem applies to trochospiral *Rhizocorallium uliarense* and the considerable lengths in planar *Rhizocorallium irregulare*. As seen from the occasional S-shaped outline of the laminae, the basic U-shape sometimes became modified.

There are different opinions on the relations between *Rhizocorallium* and *Zoophycos*. Seilacher (1967, 1986) states that there is a clear distinction, with *Rhizocorallium* having tubes that are wide with respect to spreite width (see also Häntzschel 1960), and where the U-bend remains constant; although lobate forms occur, they are built up of the same U-shaped units. *Zoophycos*, on the other hand, has narrow tubes and a composite spreite, not fixed by a certain curvature (Seilacher 1986). However, Ekdale & Lewis (1991) recently interpreted specimens of *Zoophycos* from New Zeeland to have formed by overlapping, *Rhizocorallium*-like spreiten. These authors stressed that testing is needed to decide if their model applies also to other occurrences of *Zoophycos*.

This type of burrow is typically interpreted as a feeding structure representing activity during a long period of the originator's life (Wetzel & Werner 1981).

Zoophycos isp.
Fig. 65B–C, ?D

Synonymy. – □1881 *Spirophyton velum* – Nathorst, p. 37.
□1881 *Spirophyton* cfr. *velum* – Nathorst, p. 52.

Material. – Figured slabs, RM X3340, 3341, 3348. Numerous specimens in collections of RM. Specimens from quarries at Älerud, Lugnås, found 6.5–7.0 m above the basement. Several of the specimens are from glauconitic sandstone.

Description. – Horizontal spreite with lobate outline, rimmed by a marginal area about 1 cm wide (Fig. 65B–C).

Proximal part of spreite with fairly regular U-shaped laminae, toward the distal region with laminae becoming increasingly wider and more lobate. In the proximal part, distance between the limbs of marginal burrow is about 2 cm; in the widest part about 10 cm. Marginal burrow horizontal, except at proximal part, where one arm turns sharply upwards. The marginal tube locally shows a low spreite. Endogenic traces preserved on upper surface of slab. Spreite horizontally protrusive, vertically retrusive.

Discussion. – Nathorst (1881b, pp. 37, 52) identified probably identical burrows from the Mickwitzia sandstone in Lugnås as *Spirophyton velum* (Vanuxem, 1842). Vanuxem (1842, p. 176) suggested the name *Fucoides velum* for Devonian forms from New York that looked like folds of a curtain. There is no reference to a figure associating the name, but Vanuxem (1842, p. 159) accompanies a form 'which resembles a curtain and its folds, supported at both ends' with an illustration (Vanuxem 1842, Fig. 39). This illustration was taken to be a *Fucoides velum* by Hall (1863, Fig. 3), and he included the species in the new genus, *Spirophyton*. From the available information, Nathorst's (1881b) assignment of the Mickwitzia sandstone form to *Spirophyton velum* seems reasonable. Hall (1863) found it likely that *Spirophyton velum* was part of a spiral structure and that the different species of *Spirophyton* might turn out to be different aspects of the same species. Fritel (1925) followed this line and put several related species, including those of Hall (1863), in synonymy as *Spirophyton cauda galli*. More recently, Plička (1968) placed *Fucoides velum* in synonymy with *Zoophycos circinatus* (Brongniart, 1828).As was stated by Brongniart (1828) *Fucoides circinatus* from the Lingulid sandstone at Kinnekulle is a vertical form and thus not related to *Zoophycos* (see Jensen & Bergström 1995).

The taxonomy of *Zoophycos* is confused; no consensus exists on which is its type species and how to define the multitude of species, many of which were originally described as plants (see, Taylor 1967; Simpson 1970; Fillion & Pickerill 1984). Until a thorough revision of *Zoophycos* is made from an ichnotaxonomical basis, there is little meaning in assigning the above described form to any species.

There is a close similarity between flat *Zoophycos* and horizontal *Rhizocorallium*, and both may be seen as systematic feeding or feeding–dwelling structures with a stronger parallelism of the limbs in *Rhizocorallium* as the main difference (see also discussion on *Zoophycos* (?*Rhizocorallium*) isp.). Häntzschel & Reineck (1968) reported occurrences of early Jurassic *Rhizocorallium* from Helmstedt, Germany , including some with a lobate spreite. There seems to be a gradual transition from *Zoophycos* to *Rhizocorallium*. The laminated marginal burrow in the Mickwitzia sandstone specimens is indicative of

Fig. 64. Zoophycos (?*Rhizocorallium*) isp., preserved on lower surface of medium-grained sandstone slabs. Also seen are *Gyrolithes polonicus*. Scale bars 10 mm. Hjälmsäter. Interval B (0.4–1.0 m above basement) □A. RM X3228. □B. RM X3225.

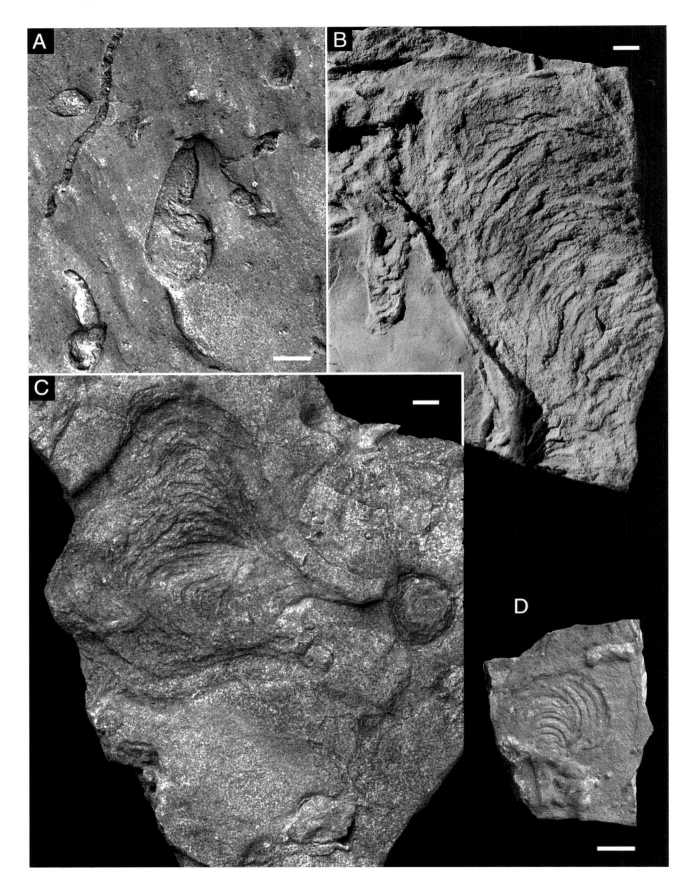

some change of level during the time of spreite formation, possibly caused by slight sediment accumulation.

Arcuate burrows

Fig. 66

Material. – Figured slabs, LO 6699t, RM X3232, 3314. Lugnås and Kinnekulle. Interval B.

Description. – Flat, gently inclined structures, in deepest part with a smooth convex rim (Fig. 66). Borders of trace steep or convex, abruptly turning into flat main part. These structures are arranged in a series with successive traces overlapping in an imbricated manner (Fig. 66C). In one specimen overlapping is very close, making individual traces difficult to distinguish (Fig. 66A). Two of the series of traces are closely associated with *Rusophycus dispar* burrows (Fig. 66A–B). Maximum widths of associated arcuate structures and *Rusophycus* for the two series are 35 and 33 mm, and 29 and 27 mm, respectively.

Interpretation. – These traces may have been made by an arthropod, probably trilobite, ploughing into sediment with its head shield. Whittington (1980) and Briggs & Whittington (1985) suggested that the middle Cambrian trilobite *Olenoides serratus*, engaged in shallow ploughing by lowering the anterior part of the head-shield into the sediment while walking forwards. Whittington (1980) stressed that during forward propulsion the tips of the walking legs would remain at a more or less fixed point in the sediment during the backward thrust, which would explain the lack of scratch marks. The trilobite either remained more or less stationary and pushed the head shield into the sediment, or made a continuous forward ploughing movement. In the latter case, the convex ridges in Fig. 66A would have resulted from a bobbing motion of the head during forward motion. The strongest indication that the arcuate structures and *Rusophycus dispar* were made by the same animal are the two associations with corresponding size of *Rusophycus* and the arcuate burrow. As proposed by Whittington (1980), the motivation behind this type of activity was probably search for prey. The associated *Rusophycus* burrows may represent instances where prey was located and dug after (Fig. 66A).

There appear to be no reports of traces with the above interpretation, though there are trilobite trace fossils with cephalic impressions. In cruzianiform specimens of *Rusophycus marginatus* Bergström & Peel, 1988, there are repeated imprints made by the head shield, leading Bergström & Peel (1988) to suggest a bobbing motion of the head shield, perhaps coordinated with leg movements. The role of the head shield in this trace is not clear, and the abundant leg marks indicate that they were the major agent in the excavation.

Rusophycus cryptolithi are burrows with head shield impressions made by trinucleid trilobites. Campbell (1975) deduced that the head shield was not used in ploughing, but that the trilobite made discrete but overlapping burrows or burrowed backwards. *Rusophycus perucca* and *R. jenningsi* also show head-shield impressions, but not of a type indicating ploughing. In these traces, the head shield probably had a subordinate importance to the legs in digging, or it was used solely to bury the trilobite in sediment. Possibly analogous traces are known as *Selenichnites rossendalensis* (Hardy, 1970). This trace fossil consists of 'lunate casts corresponding in outline to the shape of the xiphosurian prosoma, often in a

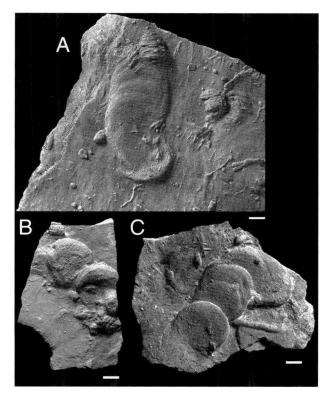

Fig. 66. Arcuate burrows possibly formed by ploughing trilobites, or laterally creeping cnidarians. Scale bars 10 mm. □A. Lugnås. RM X3342. □B. Hjälmsäter, interval B, 0.4–1 m above basement. RM X3232. □C. Lugnås. This could be a *Bergaueria sucta*. LO 6699t.

Fig. 65 (opposite page). □A. *Rhizocorallium* isp. A specimen that has weathered out at the base of a slab of fine sandstone. Also seen are meniscate burrows. Scale bar 10 mm. Trolmen. RM X3339. □B. ?*Zoophycos* isp., Scale bar 10 mm. Loose block, Älerud, Lugnås. From a level higher than 6 m. RM X3340. □C. *Zoophycos* isp. Lobate specimen in glauconitic sandstone. On the left side of the slab are two specimens of *Rosselia socialis*, the upper of which has an inclined orientation. Scale bar 10 mm. Lugnås. RM X3341. □D. Specimen intermediate between *Rhizocorallium* and *Zoophycos*, Scale bar 10 mm. Hjälmsäter. RM X3348

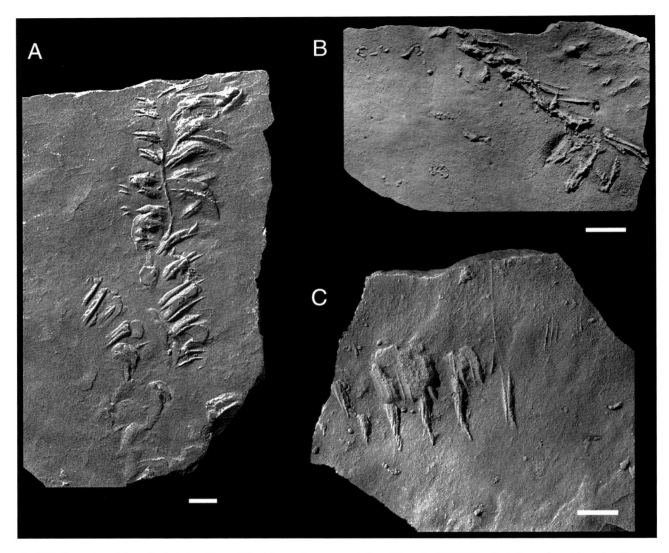

Fig. 67. Arthropod scratch marks. Scale bar 10 mm. □A. Arthropod scratch mark Type B. Lugnås. SGU 8599. □B. Arthropod scratch marks. Also seen are probable '*Hormosiroidea*' isp., Hjälmsäter, interval B. RM X3343. □C. Arthropod scratch marks. This specimen shows imprints identical to those of *Monomorphichnus* isp. B, to which it may belong. The deepest ridges are folded. Note small *Rusophycus eutendorfensis*. Hjälmsäter, interval B. RM X3344.

series, and associated with appendage and telson marks' (Hardy 1970, p. 189). The traces are also similar to those described here, by being deepest in the convex part and shallowing posteriorly (Hardy 1970, p. 189; Eagar *et al.* 1985; Romano & Whyte 1987, 1990).

The trilobite fauna of the Baltic region was during much of the Lower Cambrian dominated by olenellids, apparently endemic to the province (Ahlberg *et al.* 1986). Of these, *Kjerulfia*? *lundgreni* has an outline that fits well with that of the impression on the specimen on Fig. 66B (see Bergström 1973b, Fig. 18).

There may be other interpretations of this form. They could be laterally displaced bergauerians, making them examples of *Bergaueria sucta* (see above) or they may be inclined *Rhizocorallium* with different amounts of over-

lapping (cf. Eagar *et al.* 1985, Pl. 14B, C, E). However, the association with *Rusophycus* supports the above interpretation.

Arthropod scratch marks

Under this heading are included supposed claw imprints of arthropod origin that cannot be assigned to *Rusophycus*, *Cruziana* or *Monomorphichnus*. However, in many cases, these scratch marks are probably undertrack fallouts of the two first or represent parts of the third (Fig. 67B–C). Similar traces are commonly reported in the literature as *Diplichnites* Dawson, 1873, and generally interpreted as walking traces of arthropods, and especially trilobites (e.g., Seilacher 1955a; Crimes 1970). The use of *Diplichnites* is avoided since the taxonomic status of *Diplichnites* is uncertain (see Briggs *et al.* 1984) and the interpretation of this type of trace as walking traces is highly doubtful.

Arthropod scratch mark Type A

Fig. 42 (part)

Material. – Figured slab, SGU 8593. Lugnås. Probably interval B.

Description. – Seven pairs of elongated triangular scratch marks, each scratch mark slightly inclined to the long axis of the trace (Fig. 42). The distance between the marks of a pair is about 2 cm, except for a deeper pair where the distance is smaller. The surface of each mark is striated in its longest direction, with up to 6 (?8) ridges. Left side of pairs (on slab) are shorter; length of left and right scratch marks in a pronounced pair is 13 and 22 mm, respectively. Found together with, and in places cut trough and disrupted by, *Monomorphichnus* isp. B.

Discussion. – The similar inclination of the opposing pairs indicates that they belong to corresponding leg pairs. These markings correspond closely to isolated proverse scratch marks in *Rusophycus dispar*, and they are probably undertrack fallout of *Cruziana rusoformis*.

Arthropod scratch mark Type B

Fig. 67A

Material. – Figured slab, SGU 8599. Lugnås. Probably interval B.

Description. – Two sets of slightly superimposed markings set at angle of about 30° to each other (Fig. 67A). Both sets show asymmetry in length and direction of their left and right markings. In the second set the left markings are short and straight and the right markings long and curved. Most markings consist of paired ridges, though additional ridges are also seen. In the second set, a ridge runs between the marking sets for about three-quarters of their extent on the slab. A corresponding ridge may be present along part of the first set.

Discussion. – The asymmetry and lateral shift in this trace may indicate the influence of currents. The sharp shift in direction could be caused by a current-caused hop (cf. Briggs & Rushton 1980). The median ridge probably stems from part of the animal; in the case of a trilobite a genal spine could be considered. Another possibility is the impression of a long axial spine known to have existed in at least one olenellid in the Baltic region; *Schmidtiellus mickwitzi* (see Bergström 1973b, Fig. 9).

Acknowledgments. – First and foremost I want to thank Stefan Bengtson, who has provided guidance and support, improved the language, and given constructive criticism during all stages of the work. This work was carried out at the Palaeontological Institute, Uppsala University, were working facilities were kindly provided by Professors Richard A. Reyment during the initial, and John S. Peel during the final stage of the work. John S. Peel also improved the language and provided helpful support during the final stages of this study.

During a two months' visit at the Palaeontological Institute, Moscow, Dr. Mikhail Fedonkin gave me free access to his rich material and knowledge of Vendian fossils. This stay, and a subsequent field trip to the White Sea region lead by Dr. Fedonkin, gave invaluable first-hand experience with Vendian fossils. For information and discussions on various aspects of trace fossils and problematica, for giving access to collections and for their hospitality, I am grateful to Dr. Richard Bromley, Copenhagen; Dr. Simon Conway Morris, Cambridge; Dr. Peter Crimes, Liverpool; Drs. Antonio Liñan and Antonio Gamez, Zaragoza; Dr. Bruce Runnegar, Los Angeles, and Dr. Jim Gehling, Adelaide. Dr. Peter Crimes provided constructive critizism of manuscripts.

For arranging loans of museum material my gratitude to Professor Jan Bergström, Stockholm; Dr. Per Ahlberg, Lund; and the staff of the collections of the Swedish Geological Survey

A special thank to Jan Johansson, Sköllersta, for giving me free access to his rich collection of Mickwitzia sandstone material. For loan of material I thank Holger Buentke, Lugnås and Allan Karlsson, Medelplana.

I thank all my colleagues and friends at the Department of Historical Geology & Palaeontology, Uppsala, for good company. A special thank to Stephen W.F. Grant for many stimulating and enlightening discussions on various subjects, including palaeontology. For technical assistance my sincerest thanks to Christer Bäck and Birgit Jansson, Uppsala.

My work has been supported by funds from the Swedish Natural Science Research Council (NFR), Royal Swedish Academy of Science and Uppsala University. Publication costs were defrayed through NFR grant No. P-IP 05239-306.

References

Aceñolaza, F.G. & Manca, N. del V. 1982: *Bifungites* sp. (traza fosils) en capas del Ordovicico inferior de la region de Perchel, Quebrada de Humahuaca, provincia de Jujuy. *Ameghiniana 19*, 157–164.

Ahlberg, P. 1984: Lower Cambrian trilobites and biostratigraphy of Scandinavia. *Lund Publications in Geology 22*. 37 pp.

Ahlberg, P. 1989: Cambrian stratigraphy of the Når 1 deep well, Gotland. *Geologiska Föreningens i Stockholm Förhandlingar 111*, 137–148.

Ahlberg, P. & Bergström, J. 1993: The trilobite *Calodiscus lobatus* from the Lower Cambrian of Scania, Sweden. *Geologiska Föreningens i Stockholm Förhandlingar 115*, 331–334.

Ahlberg, P., Bergström, J. & Johansson, J. 1986: Lower Cambrian olenellid trilobites from the Baltic Faunal Province. *Geologiska Föreningens i Stockholm Förhandlingar 108*, 39–57.

Aigner, T. 1985: *Storm Depositional Systems.* Lecture Notes in Earth Sciences. Springer-Verlag. 174 pp.

Aigner, T. & Futterer, E. 1978. Kolk-Töpfe und -Rinnen (pot and gutter casts) im Muschelkalk – Anzeiger für Wattenmeer? *Neues Jahrbuch für Geologie und Paläontologie, Abhandlungen 156*, 285–304.

Aigner, T. & Reineck, H.-E. 1982: Proximity trends in modern storm sands from the Helogoland Bight (North Sea) and their implications for basin analysis. *Senckenbergiana Maritima 14*, 183–215.

Alexandrescu, G. & Brustur, T. 1987: Structures sedimentaires biogènes (trace fossils) du flysch des Carpathes Orientales. *Dări de seamă ale şedinţelor, Institutul Geologie şi Geofizică 72–73:3, Paleontologie*, 5–20.

Allen, J.R.L. 1982: *Sedimentary Structures: Their Character and Physical Basis.* Development in Sedimentology 30:I & II. Elsevier, Amsterdam.

Allen, J.R.L. 1985: Wrinkle marks: an intertidal sedimentary structure due to aseismic soft-sediment loading. *Sedimentary Geology 41*, 75–95.

Allen, J.R.L. 1987: Desiccation of mud in the temperate intertidal zone: studies from the Severn estuary and eastern England. *Philosophical Transactions of the Royal Society, London B315*, 127–156.

Aller, R.C. 1982: The Effects of macrobenthos on chemical properties of marine sediment and overlying water. *In* McCall, P.L. & Tevesz, M.J.S. (eds.): *Animal–Sediment Relations. The Biogenic Alteration odf Sediments*, 53–102. Plenum Press, New York, N.Y.

Alpert, S. 1973: *Bergaueria* Prantl (Cambrian and Ordovician), a probable actinian trace fossil. *Journal of Paleontology 47*, 919–924

[Alpert, S. 1974a: Trace fossils of the Precambrian–Cambrian succession, White–Inyo Mountains, California. Unpublished Ph.D. Thesis, University of California, Los Angeles, Cal. 162 pp.]

Alpert, S. 1974b: Systematic review of the genus *Skolithos*. *Journal of Paleontology 48*, 661–669.

Alpert, S. 1975: *Planolites* and *Skolithos* from the Upper Precambrian–Lower Cambrian White–Inyo Mountains, California. *Journal of Paleontology 49*, 508–521.

Alpert, S. 1976a: Trace fossils of the White–Inyo Mountains. *In* Moore, J.N. & Fritsche, A.E. (eds.): Depositional environments of Lower Paleozoic rocks in the White–Inyo Mountains, Inyo County, California, 43–48. *Pacific Section, Society of Economic Palaeontologists and Mineralogists Pacific Coast Paleogeography Field Guide 1.*

Alpert, S. 1976b: Trilobites and star-like trace fossils from the White–Inyo Mountains, California. *Journal of Paleontology 50*, 226–239.

Alpert, S. 1977: Trace fossils and the basal Cambrian boundary. *In* Crimes, T.P. & Harper, J.C. (eds.): *Trace Fossils 2, 1–8. Geological Journal Special Issue 9.*

Anderton, R. 1976: Tidal shelf sedimentation: an example from the Scottish Dalradian. *Sedimentology 23*, 429–458.

Angelin, N.P. 1854: *Palaeontologia Scandinavia. P.I. Crustacea Formationis Transitionis. Fasc II.* T.O. Weigel, Lipsiae.

Archer, A.W. & Maples, C.G. 1984: Trace-fossil distribution across a marine-to-nonmarine gradient in the Pennsylvanian of south-western Indiana. *Journal of Paleontology 58*, 448–466.

Astin, T.R. & Rogers, D.A. 1991: 'Subaqueous shrinkage cracks' in the Devonian of Scotland reinterpreted. *Journal of Sedimentary Petrology 61*, 850–859.

Azpeita Moros, F. 1933: Datos para el estudio paleontólogico del Flysch de la Costa Cantábrica y de algunos otros puntos de España. Boletin del Instituto Geologico y Minero de España 53, 1–65.

Badve, R.M. & Ghare, M.A. 1978: Jurassic Ichnofauna of Kutch – I. *Biovigyanam 4*, 125–140.

Baldwin, C.T. 1977: Internal structures of trilobite trace fossils indicative of an open surface furrow origin. *Palaeogeography, Palaeoclimatology, Palaeoecology 21*, 273–284.

Bandel, K. 1967: Trace fossils from two Upper Pennsylvanian sandstones in Kansas. *University of Kansas Paleontological Contributions, Paper 18*, 1–13.

Banks, N.L. 1970: Trace fossils from the late Precambrian and Lower Cambrian of Finnmark, Norway. *In* Crimes, T.P. & Harper, J.C. (eds.): *Trace Fossils*, 19–34. *Geological Journal Special Issue 3.*

Barbour, I.H. 1892: Note on new gigantic fossils. *Science 19*, 99–100.

Barwis, J.H. 1985: Tubes of the modern polychate *Diopatra cuprea* as current velocity indicators and as analogs for *Skolithos – Monocraterion*. *In* Allen Curran, H. (ed.): Biogenic structures: their use in interpreting depositional environments, 225–235. *Society of Economic Paleontologists and Mineralogists Special Publication 35.*

Bassler, R.S. 1941: A supposed jelly fish from the pre-Cambrian of the Grand Canyon. *Proceedings, United States National Museum 89:3104*, 519–522.

Baumfalk, Y.A. 1979: Heterogenous grain size distribution in tidal flat sediment casued by bioturbation activity of *Arenicola marina* (Polychaeta). *Netherlands Journal of Sea Research 13*, 428–440.

Bednarczyk, W. & Przybyłowicz, T. 1980: On development of Middle Cambrian sediments in the Gdansk Bay area. *Acta Geologica Polonica 30*, 391–415.

Bengtson, P. 1980: Orthography of geological names derived from fossil taxa. *Geologiska Föreningens i Stockholm Förhandlingar 102*, 222.

Bengtson, P. 1988: Open nomenclature. *Palaeontology 31*, 223–227.

Bergström, J. 1968: *Eolimulus*, a Lower Cambrian Xiphosurid from Sweden. *Geologiska Föreningens i Stockholm Förhandlingar 90*, 489–503.

Bergström, J. 1970: *Rusophycus* as an indication of early Cambrian age. *In* Crimes, T.P. & Harper, J.C. (eds.): *Trace Fossils*, 35–42. *Geological Journal Special Issue 3.*

Bergström, J. 1971: *Paleomerus* – merostome or merostomoid. *Lethaia 4*, 393–401.

Bergström, J. 1973a: Organization, life and systematics of trilobites. *Fossils and Strata 2*, 1–69.

Bergström, J. 1973b: Classification of olenellid trilobites and some Balto-Scandian species. *Norsk Geologisk Tidskrift 53*, 283–314.

Bergström, J. 1976: Lower Palaeozoic trace fossils from eastern Newfoundland. *Canadian Journal of Earth Sciences 13*, 1613–1633.

Bergström, J. 1981: Lower Cambrian shelly faunas and biostratigraphy in Scandinavia. *In* Taylor, M.E. (ed.): Short papers for the Second International Symposium on the Cambrian System 1981, 22–25. *U.S. Geological Survey Open-File Report 81-743.*

Bergström, J. 1989: The origin of animal phyla and the new phylum Procoelomata. *Lethaia 22*, 259–269.

Bergström, J. 1990: Precambrian trace fossils and the rise of bilaterian animals. *Ichnos 1*, 3–13.

Bergström, J. 1991: Metazoan evolution around the Precambrian–Cambrian transition. *In* Simonetta, A.M. & Conway Morris, S. (eds.): *The Early Evolution of the Metazoa and the Significance of Problematic Taxa*, 25–34. Cambridge University Press, Cambridge.

Bergström, J. & Gee, D.G. 1985: The Cambrian in Scandinavia. *In* Gee, D.G. & Sturt, B.A. (eds.): *The Caledonide Orogen – Scandinavia and related Areas*, 247–271. Wiley, New York, N.Y.

Bergström, J. & Peel, J.S. 1988. Lower Cambrian trace fossils from northern Greenland. *Grønlands Geologiske Undersøgelse, Rapport 137*, 43–53.

Billings, E. 1862: New species of fossils from different parts of the Lower, Middle and Upper Silurian rocks of Canada. *Palaeozoic Fossils 1*, 96–102. Geological Survey of Canada, Advance Sheets.

Billings, E. 1872: On some fossils from the primordial rocks of Newfoundland. *The Canadian Naturalist, New Series, 6*, 465–479.

Birkenmajer, K. & Bruton, D.L. 1971: Some trilobite resting and crawling traces. *Lethaia 4*, 303–319.

Biron, P.E. & Dutoit, J.-M. 1981: Figurations sédimentaires et traces d'activité au sol dans le Trias de la formation d'Argana et de l'Ourika (Maroc). *Bulletin du Muséum d'histoire naturelle, Paris, 4e série, 3, 1981, section C, no 4*, 399–427.

Bjerstedt, T.W. & Erickson, J.M. 1989: Trace fossils and bioturbation in peritidal facies of the Potsdam–Theresa Formations (Cambrian–Ordovician), Northwest Adirondacks. *Palaios 4*, 203–224.

Bland, B.H. 1984: *Arumberia* Glaessner & Walter, a review of its potential for correlation in the region of the Precambrian–Cambrian boundary. *Geological Magazine 121*, 625–633.

Bornemann, J.G. 1889: Über den Buntsandstein in Deutschland und seine Bedeutung für die Trias. *Beiträge Geologie, Paläontologie 1*, 1–61.

Bose, P.K. & Chaudhuri, A.K. 1990: Tide versus storm in Epeiric coastal deposition: two Proterozoic sequences, India. *Geological Journal 25*, 81–101.

Böttcher, R. 1982: Die Abu Ballas Formation (Lingula Shale) (Apt?) der Nubischen Gruppe Südwest-Ägyptens. Eine Beschribung der Formation unter besonderer Berücksichtigung der Paläontologie. *Berliner Geowissenschaftliche Abhandlungen A39*. 145 pp.

Bottjer, D.J., Droser, M.L. & Jablonski, D. 1987: Bathymetric trends in the history of trace fossils. *In* Bottjer, D.J. (ed.): *New Concepts in the Use of Biogenic Sedimentary Structures for Paleoenvironmental Interpretation*, 57–65. *Society of Economic Paleontologists and Mineralogists, Pacific Section, Volume and Guidbook 52.*

Bown, T.M. & Kraus, M.J. 1983: Ichnofossils of the alluvial Willwood Formation (Lower Eocen), Bighorn Basin, Northwest Wyoming, U.S.A. *Palaeogeography, Palaeoclimatology, Palaeoecology 43*, 95–128.

Bradshaw, M.A. 1981: Paleoenvironmental interpretations and systematics of Devonian trace fossils from the Taylor Group (lower Beacon Supergroup), Antarctica. *New Zealand Journal of Geology and Geophysics 24*, 615–652.

Brangulis, A.P. 1985: *Vend i Kembrij Latvii.* [The Vendian and Cambrian of Latvia.] 134 pp. Zinate, Riga.

Brenchley, P.J. 1985: Storm influenced sandstone beds. *Modern Geology 9*, 369–396.

Brenchley, P.J., Romano, M. & Gutierrez-Marco J.C. 1986: Proximal and distal hummocky cross-stratified facies on a wide Ordovician shelf in Iberia. *In* Knight, R.J. & McLean, J.R. (eds.): *Shelf Sands and Sandstones*, 241–255. *Canadian Society of Petroleum Geologists, Memoir 2.*

Bridge, J.S. & Droser, M.L. 1985: Unusual marginal-marine lithofacies from the Upper Devonian Catskill clastic wedge. *Geological Society of America Special Paper 201*, 143–161.

Briggs, D.E.G., Plint, A.G. & Pickerill, R.G. 1984: *Arthropleura* trails from the Westphalian of Eastern Canada. *Palaeontology 27*, 843–855.

Briggs, D.E.G. & Rushton, W.A. 1980: An arthropod trace fossil from the Upper Cambrian Festiniog Beds of North Wales and its bearing on trilobite locomotion. *Geologica et Palaeontologica 14*, 1–8.

Briggs, D.E.G., & Whittington, H.B. 1985: Modes of life of arthropods from the Burgess Shale, British Columbia. *Transactions of the Royal Society of Edinburgh 76*, 149–160.

Britton, J.C. & Morton, B. 1989: *Shore Ecology of the Gulf of Mexico.* 387 pp. University of Texas Press, Austin.

Bromley, R.G. 1990: *Trace Fossils: Biology and Taphonomy.* 280 pp. Unwin & Hyman, London.

Bromley, R.G. & Asgaard, U. 1979: Triassic freshwater ichnocoenoses from Carlsberg Fjord. *Palaeogeography, Palaeoclimatology, Palaeoecology 28*, 39–80.

Bromley, R.G. & Frey, R.W. 1974: Redescription of the trace fossil *Gyrolithes* and taxonomical evaluation of *Thalassinoides, Ophiomorpha* and *Spongeliomorpha. Bulletin of the Geological Society of Denmark 23*, 311–335.

Bromley, R.G. & Hanken, N.M. 1991: The growth vector in trace fossils:examples from the Lower Cambrian of Norway. *Ichnos 1*, 261–276.

Brongniart, M.A. 1828: *Histoire des végétaux fossiles ou recherches botaniques et géologiques sur les végétaux renfermés dans les diverses couches du globe.* 136 pp. Dufour & d'Ocagne, Paris. [Plates published 1837.]

Bruun-Petersen, J. 1973: 'Conical structures' in the Lower Cambrian Balka Sandstone, Bornholm (Denmark), and in the lower Devonian Coblenz Sandstone, Marburg (Western Germany). *Neues Jahrbuch für Geologie und Paläontologie, Monatshefte 1973:9*, 513–528.

Bryant, I.D. & Pickerill, R.K. 1990: Lower Cambrian trace fossils from the Buen Formation of central North Greenland: preliminary observations. *In* Peel, J.S. (ed.): *Lower Cambrian Trace Fossils from Greenland*, 44–62. *Grønlands Geologiske Undersøgelse, Rapport 147.*

Buatois, L.A. & Mángano, M.G. 1993: The ichnotaxonomic status of *Plangtichnus* and *Treptichnus. Ichnos 2*, 217–224.

Byers, C.W. 1982: Geological significance of marine biogenic sedimentary structures. *In* McCall, P.L. & Tevesz, M.J.S. (eds.): *Animal–Sediment Relations. The Biogenic Alteration of Sediments*, 221–256. Plenum Press, New York, N.Y.

Campbell, K.S.W. 1975: The functional morphology of *Cryptolithus. Fossils and Strata 4*, 65–86.

Chamberlain, C.K. 1971: Morphology and ethology of trace fossils from the Ouchita Mountains, southeast Oklahoma. *Journal of Paleontology 45*, 212–246.

Chiplonkar, G.W. & Badve, R.M. 1970: Trace fossils from the Bagh Beds. *Journal of the Palaeontological Society of India 14*, 1–10.

Clark, R.B. 1964: *Dynamics in Metazoan Evolution. The Origin of the Coelom and Segments.* 313 pp. Clarendon Press, Oxford.

Clausen, C.K. & Vilhjálmsson, M. 1986: Substrate control of Lower Cambrian trace fossils from Bornholm, Denmark. *Palaeogeography, Palaeoclimatology, Palaeoecology 56*, 51–68.

Cloud, P.E. 1968: Pre-Metazoan evolution and the origins of the Metazoa. *In* Drake, E.T. (ed.): *Evolution and Environment*, 1–72. Yale University Press, New Haven, Conn.

Cloud, P.E. 1973: Pseudofossils: a plea for caution. *Geology 1*, 123–127.

Cornish, F.G. 1986: The trace-fossil *Diplocraterion*; evidence of animal–sediment interactions in Cambrian tidal deposits. *Palaios 1*, 478–491.

Crimes, T.P. 1968: *Cruziana*: a stratigraphically useful trace fossil. *Geological Magazine 105*, 360–364.

Crimes, T.P. 1970: Trilobite tracks and other trace fossils from the Upper Cambrian of North Wales. *Geological Journal 7*, 47–68.

Crimes, T.P. 1975a: The stratigraphical significance of trace fossils. *In* Frey, R.W. (ed.): *The study of Trace Fossils*, 109–130. Springer-Verlag, Berlin.

Crimes, T.P. 1975b: The production and preservation of trilobite resting and furrowing traces. *Lethaia 8*, 35–48.

Crimes, T.P. 1987: Trace fossils and correlation of late Precambrian and early Cambrian strata. *Geological Magazine 124*, 97–119.

Crimes, T.P. 1989: Trace fossils. *In* Cowie, J.W & Brasier, M.D. (eds.): *The Precambrian–Cambrian Boundary*, 166–185. *Oxford Monographs on Geology and Geophysics 12.* Clarendon Press, Oxford.

Crimes, T.P. 1992: Changes in the trace fossil biota across the Proterozoic–Phanerozoic boundary. *Journal of the Geological Society, London 149*, 637–646.

Crimes, T.P. 1994: The period of early evolutionary failure and the dawn of evolutionary success: the record of biotic changes across the Precambrian–Cambrian boundary. *In* Donovan, S.K. (ed.): *The Palaeobiology of Trace Fossils*, 105–133. John Wiley & Sons, Chichester.

Crimes, T.P. & Anderson, M.M. 1985: Trace fossils from late Precambrian – early Cambrian strata of southeastern Newfoundland (Canada): Temporal and Environmental implications. *Journal of Paleontology 59*, 310–343.

Crimes, T.P. & Crossley, J.D. 1991: A diverse ichnofauna from Silurian flysch of the Aberystwyth Grits Formation, Wales. *Geological Journal 26*, 27–64.

Crimes, T.P. & Fedonkin, M.A. 1994: Evolution and dispersal of deepsea traces. *Palaios 9*, 74–83.

Crimes, T.P. & Marcos, A. 1976: Trilobite traces and the age of the lowest part of the Ordovician reference section for NW Spain. *Geological Magazine 113*, 349–356.

Crimes, T.P., Legg, I., Marcos, A. & Arboleya, M. 1977: ?Late Precambrian– low Lower Cambrian trace fossils from Spain. *In* Crimes, T.P. & Harper, J.C. (eds.): *Trace Fossils 2*, 91-138. *Geological Journal Special Issue 9.*

Dahmer, G. 1937: Lebenspuren aus dem Taunusquartzit und den Siegener Schichten (Unterdevon). *Jahrbuch der Preussischen Geologischen Landesanstalt 57*, 523–539.

D'Alessandro, A. & Bromley, R.G. 1987: Meniscate trace fossils and the *Muensteria–Taenidium* problem. *Palaeontology 30*, 743–763.

Dam, G. 1990a: Taxonomy of trace fossils from the shallow marine Lower Jurassic Neill Klinter Formation, East Greenland. *Bulletin of the Geological Society, Denmark 38*, 119–144.

Dam, G. 1990b: Palaeoenvironmental significance of trace fossils from the shallow marine Lower Jurassic Neill Klinter Formation, East Greenland. *Palaeogeography, Palaeoclimatology, Palaeoecology 79*, 221–248.

Dawson, J.W. 1873. Impressions and footprints of aquatic animals and imitative markings on Carboniferous rocks. *American Journal of Science and Arts 105*, 16–24

Dott, R.H. & Burgeois, J. 1982: Hummocky stratification: Significance of its variable bedding sequences. *Bulletin of the Geological Society of America 93*, 663–680.

Dott, R.H. & Byers, C.W. 1981 (convenors): SEPM Reserach conference on modern shelf and ancient cratonic sedimentation – the ortho-quartzite–carbonate suite revisited. *Journal of Sedimentary Petrology 51*, 329–347.

Dott, R.H., Byers, C.W., Fielder, G.W., Stenzel, S.R. & Winfree, K.E. 1986: Aeolian to marine transition in Cambro-Ordovician cratonic sheet sandstones of the northern Mississippi valley, U.S.A. *Sedimentology 33*, 345–367.

Droser, M.L. 1991: Ichnofabric of the Paleozoic *Skolithos* ichnofacies and the nature and distribution of *Skolithos* piperock. *Palaios 6*, 316–325.

Droser, M.L. & Bottjer, D.J. 1986: A semiquantitative field classification of ichnofabric. *Journal of Sedimentary Petrology 56*, 558–559.

Droser, M.L. & Bottjer, D.J. 1989: Ichnofabric of sandstones deposited in high-energy nearshore environments: measurement and utilization. *Palaios 4*, 598–604.

Durand, J. 1984: Ichnocoenose du Grès Armoricain (Ordovicien Inférieur du massif Armoricain) dans leur contexte sédimentologique. *Geobios, Mémoire Spécial 8*, 189–197.

Durand, J. 1985a: Le Grès Armoricain. Sédimentologie – Trace fossiles. Milieux de dépôt. *Memoires et Documents du Centre Armoricain d'Etude Structurale des Socles 3*, 150 pp.

Durand, J. 1985b: Les traces fossiles indicateurs paléobiologiques de milieux: un example dans l'Ordovicien armoricain. *Bulletin du Muséum d'Histoire Naturelle, Paris, 4e Série, 7, 1985, Section C, No 3*, 215–227.

Dżułyński, S. & Sanders, J.E. 1962: Current marks on firm mud bottoms. *Transactions of the Connecticut Academy of Arts and Sciences 42*, 57–96.

Dżułyński, S. & Żak, C. 1960: Srodowisko sedymentacyjne piaskowców kambryjskich z Wisniówki i ich stosunek do facji fliszowej. *Rocznik Polskiego Towarzystwa Geologicznego 30*, 213–241.

Eagar, R.M.C., Baines, J.G., Hardy, P.G., Okolo, S.A. & Pollard, J.E. 1985: Trace fossil assemblages and their occurrence in Silesian (Mid-Carboniferous) deltaic sediments of the central Pennine basin, England. *In* Allen Curran, H. (ed.): *Biogenic Structures: Their Use in Interpreting Depositional Environments*, 99–149. *Society of Economic Paleontologists and Mineralogists Special Publication 35*.

Ekdale, A.A. & Lewis, D.W. 1991: The New Zealand *Zoophycos* revisited: morphology, ethology, and paleoecology. *Ichnos 1*, 183–194.

Eklund, J. 1961: Berggrunden. Kumlas urtid och framtid. *In* Samzelius, J. (ed.): *Kumlabygden. Forntid – Nutid – Framtid*, 11–198. Kumla.

Eklund, K. 1990: Lower Cambrian acritarch stratigraphy of the Bårstad 2 core, Östergötland, Sweden. *Geologiska Föreningens i Stockholm Förhandlingar 112*, 19–44.

Elliot, R.E. 1985: An interpretation of the trace fossil *Cochlichnus kochi* (Ludwig) from the East Pennine Coalfield of Britain. *Proceedings of the Yorkshire Geological Society 45*, 183–187.

Emmons, E. 1844: *The Taconic System; Based on Observations in New York, Massachusetts, Maine, Vermont, and Rhode Island.* 68 pp. Caroll & Cook, Albany, N.Y. [Not seen.]

Farmer, J., Vidal, G., Moczydłowska, M, Strauss, H., Ahlberg, P. & Siedlecka, A. 1992: Ediacaran fossils from the Innerelv Member (late Proterozoic) of the Tanafjorden area, northeastern Finnmark. *Geological Magazine 129*, 181–195.

Fedonkin, M. 1977: Precambrian–Cambrian ichnocoenoses of the east European platform. *In* Crimes, T.P. & Harper, J.C. (eds.): *Trace Fossils 2*, 183–194. *Geological Journal Special Issue 9*.

Fedonkin, M.A. 1978: Novoe mestonakhozhdenie metazoa v vende Zimnego berega. [New occurence of Vendian non-skeletal metazoa from the Winter coast.] *Doklady Akademii Nauk SSSR 239*, 1423–1426.

Fedonkin, M.A. 1979: Paleoikhnologiya dokembriya i rannego kembriya. [Paleoichnology of the Precambrian and early Cambrian.] *In* Sokolov, B.S. (ed.): *Paleontologiya Dokembriya i Rannego Kembriya*, 183–192. Nauka, Leningrad.

Fedonkin, M.A. 1980: Iskopaemye sledy dokembrijskikh Metazoa. [Trace fossils of Precambrian Metazoa.] *Izvestiya Akademii Nauk SSSR, Seriya Geologicheskaya 1980:1*, 39–46.

Fedonkin, M.A. 1981: Belomorskaja biota venda. [The Vendian White-Sea biota.] 100 pp. *Trudy Akademii Nauk SSSR 342,*

Fedonkin, M.A. 1983: Besskeletnaya fauna podol'skogo pridnestrov'ya. [Non-skeletal fauna of Podolia (Dniestr River Valley).] *In* Velikanov, V.A., Aseeva, M.A. & Fedonkin, M.A.: *Vend Ukrainy*, 128–139. Naukova Dumka, Kiev.

Fedonkin, M.A. 1985: Paleoikhnologiya vendskikh Metazoa. [Paleoichnology of the Vendian Metazoa.] *In* Sokolov, B.S. & Ivanovskij, A.B. (eds.): *Vendskaya Sistema 1. Paleontologiya*, 112–117. Nauka, Moskva.

Fedonkin, M.A. & Runnegar, B.N. 1992: Proterozoic Metazoan Trace Fossils. *In* Schopf, J.W. & Klein, C. (eds.): *The Proterozoic Biosphere: a Multidisciplinary Study*, 389–395. Cambridge University Press, Cambridge.

Fenton, C.L. & Fenton, M.A. 1937: Trilobite 'nests' and feeding burrows. *The American Midland Naturalist 18*, 446–451.

Fillion, D. 1989: Les critères discriminants à l'interieur du triptyque *Palaeophycus–Planolites–Macaronichnus*. Essai du synthèse d'un usage critique. *Comptes Rendus de Acadèmie des Sciences 309, Sèrie II*, 169–172.

Fillion, D. & Pickerill, R.K. 1984: Systematic ichnology of the Middle Ordovician Trenton Group, St. Lawrence Lowland, eastern Canada. *Maritime Sediments and Atlantic Geology 20*, 1–41.

Fillion, D. & Pickerill, R.K. 1990a: Ichnology of the Upper Cambrian? to Lower Ordovician Bell Island and Wabana groups of eastern Newfoundland, Canada. *Palaeontographica Canadiana 7*. 119 pp.

Fillion, D. & Pickerill, R.K. 1990b: Comments on 'Substrate control of Lower Cambrian trace fossils from Bornholm, Denmark'. *Palaeogeography, Palaeoclimatology, Palaeoecology 80*, 345–350.

Firtion, F. 1958: Sur la présence d'ichnites dans le Portlandien de l'Ile d'Oléron (Charente maritime). *Annales Universitatis Saraviensis – Scientia 7*, 107–112.

Fischer, P. & Paulus, B. 1969: Spurenfossilien aus den Oberen Nohn-Schichten der Blankenheimer Mulde (Eifelium, Eifel). *Senckenbergiana Lethaea 50*, 81–101.

Fischer, W.A. 1978: The habitat of the early vertebrates: trace and body fossil evidence from the Harding Formation (Middle Ordovician), Colorado. *The Mountain Geologist 15*, 1–26.

Fitch, A. 1850: A historical, topographical and agricultural survey of the County of Washington. *Transactions New York Agricultural Society 9*, 753–944.

Frey, R.W. & Chowns, T.M. 1972: Trace fossils from the Ringgold road cut (Ordovician and Silurian), Georgia. *In* Chowns, T.M. (ed.): *Sedimentary Environments in the Paleozoic Rocks of Northwest Georgia*, 25–55. *Georgia Geological Survey Guidebook 11*.

Frey, R.W. & Howard, J.D. 1985: Trace fossils from the Panther Member, Star Point Formation (Upper Cretaceous), Coal Creek Canyon, Utah. *Journal of Paleontology 59*, 370–404.

Frey, R.W. & Seilacher, A. 1980: Uniformity in marine invertebrate ichnology. *Lethaia 13*, 183–207.

Frey, R.W. & Wheatcroft, R.A. 1989. Organism–substrate relations and their impact on sedimentary petrology. *Journal of Geological Education 37*, 261–279.

Fritel, P.H. 1925: Végétaux paléozoiques et organismes problématiques de l'Ouedai. *Société Géologique de France, Bulletin, Série 4, 25*, 33–48.

Fritz, W.H. 1980: International Precambrian–Cambrian Boundary Working Group's 1979 field study to Mackenzie Mountains, Northwest Territories, Canada. *Geological Survey of Canada, Paper 80-1A*, 41–45.

Fritz, W.H. & Crimes, T.P. 1985: Lithology, trace fossils, and correlation of Precambrian–Cambrian Boundary beds, Cassiar Mountain, north-central British Columbia. *Geological Survey of Canada Paper 83-13*. 24 pp.

Fürsich, F.T. 1974a: On *Diplocraterion* Torell 1870 and the significance of morphological features in vertical, spreiten-bearing, U-shaped trace fossils. *Journal of Paleontology 48*, 952–962.

Fürsich, F.T. 1974b: Ichnogenus *Rhizocorallium*. *Paläontologische Zeitschrift 48*, 16–28.

Fürsich, F.T. 1981: Invertebrate trace fossils from the Upper Jurassic of Portugal. *Comunicações dos serviços geológicos de Portugal 67*, 153–168.

Fürsich, F.T. & Bromley, R.G. 1985: Behavioural interpretation of a rosetted spreite trace fossil: *Dactyloidites ottoi* (Geinitz). *Lethaia 18*, 199–207.

Fürsich, F.T. & Mayr, H. 1981: Non-marine *Rhizocorallium* (trace fossil) from the Upper Freshwater Molasse (Upper Miocene) of southern Germany. *Neues Jahrbuch für Geologie und Paläontologie, Monatshefte 1981:6*, 321–333.

Gaillard, C. 1980: *Megagyrolithes ardescensis* n.gen., n.sp., trace fossile nouvelle du Valanginien d'Ardeche (France). *Geobios 13*, 465–469.

Garcia-Ramos, J.C.M. 1976: Morfologia de trazas fosiles en dos aflorimientos de «Arenisca de Naranco» (Devonico Medio) de Asturias (NW. de España). *Trabajos de Geologica 8*, 131–171.

Gehling, J.G. 1991: The case for Ediacaran fossil roots to the metazoan tree. *Geological Society of India Memoir 20*, 181–224.

Gernant, R.E. 1972: The paleoenvironmental significance of *Gyrolithes* (Lebensspur). *Journal of Paleontology 46*, 735–741.

Geyer, G. & Uchman, A. 1995: Ichnofossil assemblages from the Nama Group (Neoproterozoic–Lower Cambrian) in Namibia and the Proterozoic–Cambrian boundary problem revisited. *In* Geyer, G. & Landing, E. (eds.): *Morocco '95. The Lower Cambrian standard of Western Gondwana*, 175–202. Beringeria Special Issue 2.

Gibson, G. 1989: Trace fossils from the late Precambrian Carolina Slate Belt, south-central North Carolina. *Journal of Paleontology 63*, 1–10.

Glaessner, M.F. 1969: Trace fossils from the Precambrian and basal Cambrian. *Lethaia 2*, 369–393.

Glaessner, M.F. 1984: *The Dawn of Animal Life*. Cambridge University Press, Cambridge. 244 pp.

Goldring, R. 1962: The trace fossils of the Baggy Beds (Upper Devonian) of North Devon. *Paläontologische Zeitschrift 36*, 232–251.

Goldring, R. 1971: Shallow-water sedimentation as illustrated in the Upper Devonian Baggy Beds. *Memoirs of the Geological Society of London 5*. 80 pp.

Goldring, R. 1985: The formation of the trace fossil *Cruziana*. *Geological Magazine 122*, 65–72

Goldring, R. & Seilacher, A. 1971: Limulid undertracks and their sedimentological implications. *Neues Jahrbuch für Geologie und Paläontologie, Abhandlungen 137*, 422–442.

Goodwin, P.W. & Anderson, E.J. 1974: Associated physical and biogenic structures in environmental subdivision of a Cambrian tidal sand body. *Journal of Geology 82*, 779–794.

Gureev, Yu.A. 1983: O novoj forme iskopaemykh sledov iz nizhnebaltijskikh otlozhenij podol'skogo pridnestrov'ya. [On a new form of trace fossil from lower Baltic deposits in Podolia, Dniestr Valley.] *Paleontologicheskij Sbornik, Lvov 1983:20*, 70–73.

Gureev, Yu.A. 1984: Bioglifi fanerozojs'kogo vidu o vendi podoillja ta ikh stratigrafitjne znatjennja. [Bioglyphs of Phanerozoic appearance from the Vendian of Podolia and their stratigraphich significance.] *Doklady Akademija Nauk USSR, Serija B 1984:4*, 5–8.

Gureev, Yu.A. 1988: Besskeletnaya fauna venda. [Non-skeletal Vendian fauna.] *In* Ryabenko, V.A. (ed.): *Biostratigrafiya i Paleogeograficheskie Rekonstruktsii Dokembriya Ukrainy [Biostratigraphy and Paleogeographic Reconstructions of the Precambrian of Ukraine]*, 65–81. Naukova Dumka, Kiev.

Gureev, Yu.A., Velikanov, V.A. & Ivanchenko, V.Ya. 1985: Besskelatnaya fauna v otlozheniyakh baltijskoi i berezhkovskoj serij podolii. [Non-skeletal fauna from deposits of the Baltic and Berezhkovsky series of Podolia.] *Doklady Akademija Nauk USSR, Serija B 1985:6*, 10–13.

Hadding, A. 1927: The pre-Quaternary sedimentary rocks of Sweden II. The Paleozoic and Mesozoic conglomerates of Sweden. *Meddelanden från Lund Geologisk-Mineralogiska Institution 32*, 42–171.

Hadding, A. 1929: The pre-Quaternary sedimentary rocks of Sweden. III. The Paleozoic and Mesozoic sandstones of Sweden. *Meddelanden från Lunds Geologisk-Mineralogiska Institution 41*. 287 pp.

Hagenfeldt, S.E. 1989: Lower Cambrian acritarchs from the Baltic Depression and south-central Sweden, taxonomy and biostratigraphy. *Stockholm Contributions in Geology 41*, 1–176.

Hagenfeldt, S.E. 1994: The Cambrian File Haidar and Borgholm Formations in the Central Baltic and south central Sweden. *Stockholm Contributions in Geology 43*, 69–110.

Hakes, W.G. 1976: Trace fossils and depositional environment of four clastic units, Upper Pennsylvanian Megacyclothems, Northeast Kansas. *The University of Kansas Paleontological Contributions, Article 63*. 46 pp.

Hall, J. 1852: *Palaeontology of New York. Volume I. Containing Descriptions of the Organic Remains of the Lower Devonian System (Equivalent in Part to the Middle Silurian Rocks of Europe)*. 362 pp. Benthuysen, Albany, N.Y.

Hall, J. 1863: Observations upon some spiralgrowing fucoidal remains of the Paleozoic rocks of New York. *New York State Cabinet, 16th Annual Report*, 76–83.

Hall, J. 1886: Note on some obscure organisms in the roofing slate of Washington County, New York. *New York State Museum Natural History 39th Annual Report*, 160.

Hallam, A. 1975: Preservation of trace fossils. *In* Frey, R.W. (ed.): *The Study of Trace Fossils*, 55–63. Springer, Berlin.

Hallam, A. 1981: *Facies Interpretation and the Stratigraphic Record*. 291 pp. Freeman, Oxford.

Hallam, A. & Swett, K. 1966: Trace fossils from the Lower Cambrian Pipe Rock of the north-west Highlands. *Scottish Journal of Geology 2*, 101–106.

Hamberg, L. 1991: Tidal and seasonal cycles in a Lower Cambrian shallow marine sandstone (Hardeberga Fm.) Scania, Southern Sweden. *In* Smith D.G., Reinson, G.E., Zaitlin, B.A. & Rahmani, R.A. (eds.): *Clastic Tidal Sedimentology*, 255–274. Canadian Society of Petroleum Geologists Memoir 16.

Han, Y. & Pickerill, R.K. 1994: *Phycodes templus* isp. nov. from the Lower Devonian of northwestern New Brunswick, eastern Canada. *Atlantic Geology 30*, 37–46.

Häntzschel, W. 1934: Schraubenformige und spiralige Grabgänge in turonen Sandsteinen des Zittauer Gebirges. *Senckenbergiana 16*, 313–324.

Häntzschel, W. 1935: *Xenohelix saxonica* n.sp. und ihre Deutung. *Senckenbergiana 17*, 105–108.

Häntzschel, W. 1960: Spreitenbauten (*Zoophycos* Massal.) im Septarienton Nordwest-Deutschlands. *Mitteilungen aus dem Geologischen Staatsinstitut in Hamburg 29*, 95–100.

Häntzschel, W. 1964: Die Spuren-Fauna, bioturbate Texturen und Marken in unterkambrischen Sandstein-Geschieben Norddeutschlands und Schwedens. *Der Aufschluss, Sonderhefte 14*, 88–102.

Häntzschel, W. 1975: Trace fossils and problematica. *In* Teichert, C. (ed.): *Treatise on Invertebrate Paleontology, Part W. Miscellanea, Supplement 1*. 269 pp. Geological Society of America and University of Kansas Press, Lawrence.

Häntzschel, W. & Reineck, H.-E. 1968: Fazies-Untersuchungen im Hettangium von Helmstedt (Niedersachsen). *Mitteilungen aus dem Geologischen Staatsinstitut in Hamburg 37*, 5–39.

Hardy, P.G. 1970: New Xiphosurid trails from the upper Carboniferous of Northern England. *Palaeontology 13*, 188–190.

Hecker, R.F. 1930: K nakhodke *Rhizocorallium* v volkhovskom devone. [On the discovery of *Rhizocorallium* from Devonian at the Volkov River.] *Ezhegodnik Russkogo Paleontologicheskogo Obshchestva 8*, 150–156.

Hecker, R.F. 1980: Sledy bespozvonochnykh i stigmarii v morskikh otlozheniyakh nizhnego karbona moskovskoj sineklizy. [Invertebrate traces and stigmaria from Lower Carboniferous marine deposits of the Moscow basin.] *Trudy Paleontologicheskogo Instituta Akademii Nauk SSSR 178*. 76 pp.

Hecker, R.F. 1983: Tafonomicheskie i ekologicheskie osobennosti fauny i flory glavnogo devonskogo polya. [Tafonomical and ecological peculiarities of the fauna and flora of the main Devonian areas.] *Akademiya Nauk SSSR, Trudy Paleontologicheskogo Instituta 190*. 142 pp.

Hecker, R.F., Osipova, A.I. & Bel'skaya, T.N. 1962: *Ferganskij Zaliv Paleogenovogo Morya Srednej Azii. Ego Istoriya, Osadki, Fauna, Flora, Usloviya ikh Obitaniya i Razvitie. Kniga 2 [The Fergana bay of the middle Asian Paleogene sea. Its history, fauna, flora, living conditions and development. Book 2]*. 332 pp. Moscow.

Hedberg, H.D. (ed.) 1976: *International Stratigraphic Guide*. 200 pp. John Wiley & Sons, New York, N.Y.

Hedström, H. 1923: Remarks on some fossils from the diamond boring at the Visby Cement Factory. *Sveriges Geologiska Undersökning C 314*, 1–27.

Heer, O. 1877: *Flora Fossilis Helvetiae. Die vorweltliche Flora der Schweiz*. 182 pp. Würster, Zürich.

Heinberg, C. 1973: The internal structure of the trace fossils *Gyrochorte* and *Curvolithus*. *Lethaia 6*, 227–238.

Heinberg, C. & Birkelund, T. 1984: Trace-fossil assemblages and basin evolution of the Vardekløft Formation (Middle Jurassic, central East Greenland). *Journal of Paleontology 58*, 362–397.

Hertweck, G. 1968: Die Biofazies. *In* Reineck, H.-E, Dörjes, J., Gadow, S. & Hertweck, G. Sedimentologie, Faunenzonierung und Faziesabfolge vor der Ostküste der inneren Deutschen Bucht. *Senckenbergiana Lethaea 49*, 284–296.

Hertweck, G. 1972: Georgia Coastal Region, Sapelo Isalnd, U.S.A.: Sedimentology and biology. V. Distribution and environmental significance of lebensspuren and in-situ skeletal remains. *Senckenbergiana Maritima 4*, 125–167.

Hesselbo, S.P. 1988: Trace fossils of Cambrian aglaspidid arthropods. *Lethaia 21*, 139–146.

Hessland, I. 1955: Studies on the lithogenesis of the Cambrian and basal Ordovician of the Böda Hamn sequence of strata. *Bulletin of the Geological Institution of the University of Uppsala 35*, 35–109

Hitchcock, E. 1858: *Ichnology of New England. A Report on the Sandstone of the Connecticut Valley, Especially its Footprints*. White, Boston, Mass.

Hofmann, H.J. 1971: Precambrian fossils, pseudofossils, and problematica in Canada. *Geological Survey of Canada Bulletin 189*. 146 pp.

Hofmann, H.J. 1979: Chazy (Middle Ordovician) trace fossils in the Ottawa – St. Lawrence lowlands. *Geological Society of Canada Bulletin 321*, 27–59.

Hofmann, H.J. 1990: Computer simulation of trace fossils with random patterns and the use of goniograms. *Ichnos 1*, 15–20.

Hofmann, H.J. & Patel, I.M. 1989: Trace fossils from the type 'Etcheminian Series' (Lower Cambrian Ratcliffe Brook Formation), Saint John area, New Brunswick, Canada. *Geological Magazine 126*, 139–157.

Högbom, A.G. 1925: A problematic fossil from the Lower Cambrian of Kinnekulle. *Bulletin of the Geological Institution of the University of Upsala 19*, 215–222.

Högbom, A.G. & Ahlström, N.G. 1924: Über die subkambrische Landfläche am Fusse vom Kinnekulle. *Bulletin of the Geological Institution of the University of Upsala 19*, 55–88.

Holm, G. 1901: Kinnekulles berggrund. *In* Holm, G. & Munthe, H.: *Kinnekulle, dess Geologi och den Tekniska Användningen av dess Bergarter*, 1–76. *Sveriges Geologiska Undersökning C 172*.

Holmer, L.E. 1989: Middle Ordovician phosphatic inarticulate brachiopods from Västergötland and Dalarna, Sweden. *Fossils and Strata 26*. 172 pp.

Holst, N.O. 1893: Bidrag till kännedomen om lagerföljden inom den kambriska sandstenen. *Sveriges Geologiska Undersökning C 130*. 17 pp.

Howard, R.W. & Frey, J.D. 1975: Estuaries of the Georgia Coast, U.S.A.: Sedimentology and Biology. II. Regional Animal–Sediment Characteristics of Georgia Estuaries. *Senckenbergiana Maritima 7*, 33–103.

Howard, J.D. & Frey, R.W. 1984: Characteristic trace fossils in nearshore to offshore sequences, Upper Cretaceous of east-central Utah. *Canadian Journal of Earth Sciences 21*, 200–219.

Howell, B.F. 1946: Silurian *Monocraterion clintonense* burrows showing the aperture. *Bulletin of the Wagner Free Institute of Science 21*, 29–37.

Huene, F. von 1904: Geologische Notizen aus Oeland und Dalarne, sowie über eine Meduse aus dem Untersilur. *Zentralblatt für Mineralogie, Geologie und Paläontologie 1904*, 450–461.

Jablonski, D., Sepkoski, J.J., Bottjer, D.J. & Sheehan, P.M. 1983: Onshore–offshore patterns in the evolution of Phanerozoic shelf communities. *Science 222*, 1123–1125.

Jenkins, R.J.F., Ford, C.H. & Gehling, J.G. 1983: The Ediacaran Member of the Rawnsley Quartzite: the context of the Ediacaran assemblage (Late Precambrian, Flinders Range). *Journal of the Geological Society of Australia 29*, 101–119.

Jenkins, R.J.F., Plummer, P.S. & Moriarty, K.C. 1981: Late Precambrian pseudofossils from the Flinders Ranges, South Australia. *Transactions, Royal Society of South Australia 105*, 67–83.

Jensen, S. 1990: Predation by early Cambrian trilobites on infaunal worms – evidence from the Swedish Mickwitzia Sandstone. *Lethaia 23*, 29–42.

Jensen, S. & Bergström, J. 1995: The trace fossil *Fucoides circinatus* Brongniart, 1828, from its type area, Västergötland, Sweden. *GFF 117*, 207–210.

Jensen, S. & Grant, S.W.F. 1992: Stratigraphy and paleobiology of *Kullingia* and trace fossils from the late Proterozoic to Cambrian of northern Sweden. *Geological Society of America Annual Meeting, 1992, Abstracts with Programs*, A114.

Kalm, P. 1746: *Pehr Kalms Wästgöta och Bohuslänska Resa*. New Edition, Wahlström and Widstrand, Stockholm (1977).

Kayser, F.H.E. 1872: Neue Fossilien aus dem rheinischen Devon. *Zeitschrift der Deutschen Geologischen Gesellschaft 24*, 691–700.

Keij, A.J. 1965: Miocene Trace Fossils from Borneo. *Paläontologische Zeitschrift 39*, 220–228.

Kennedy, W.J. 1967: Burrows and surface traces from the Lower Chalk of Southern England. *British Museum of Natural History (Geology) Bulletin 15*, 127–167.

Kern, J.P. 1980: Origin of trace fossils in Polish Carpathian flysch. *Lethaia 13*, 347–362.

Kilpper, K. 1963: *Xenohelix* Mansfield 1927 aus der miozänen Niederrheinischen Braunkohlenformation. *Paläontologische Zeitschrift 36*, 55–58.

Kirjanov, V.V. 1968: Paleontologicheskie ostatki i stratigrafiya otlozhenij baltijskoj serii Volyn–Podoli. *In Paleontologiya i stratigrafiya nizhnego paleozoya Volyno-Podoli*, 5–25. Naukova Dumka, Kiev.

Kitchell, J.A. 1979: Deep-sea foraging pathways: An analysis of randomness and resource exploitation. *Paleobiology 5*, 107–125.

Klein, G. de V. 1977: *Clastic Tidal Facies*. 149 pp. Continuing Education Publication Company, Champaign, Ill.

Kotake, N. 1989: Paleoecology of the *Zoophycos* producer. *Lethaia 22*, 327–341.

Kowalski, W.R. 1978: Critical analysis of Cambrian ichnogenus *Plagiogmus* Roedel, 1929. *Rocznik Polskiego Towarzystwa Geologicznego 48*, 333–344.

Książkiewicz, M. 1968: O nekotorych problematykach z fliszu karpat Polskich (Czesc III). *Rocznik Polskiego Towarzystwa Geologicznego 38*, 3–17.

Książkiewicz, M. 1977: Trace fossils in the flysch of the Polish Carpathians. *Palaeontologia Polonica 36*, 208 pp.

Landing, E, Narbonne, G.M., Benus, A.P & Anderson, A.M. 1988: Faunas and depositional environments of the Upper Precambrian through Lower Cambrian, southeastern Newfoundland. *In* Landing, E., Narbonne, G.M. & Myrow, P. (eds.): Trace fossils, small shelly fossils and the Precambrian–Cambrian boundary – proceedings, 18–52. *New York State Museum Bulletin 463*.

Lebescone, P. 1883: Les *Cruziana* et *Rysophycus* connu sous le nom général de bilobites. Sont-ils des végétaux ou des traces d'animaux. *In Ouvres Posthumes de Marie Rouault,…publies par les soins de P. Lebesconte*, 61–73. Oberthur, Rennes, Paris.

Leckie, D. 1988: Wave-formed coarse-grained ripples and their relationship to hummocky cross-stratification. *Journal of Sedimentary Petrology 58*, 607–622.

Legg, I.C. 1985: Trace fossils from a Middle Cambrian deltaic sequence, north Spain. *In* Allen Curran, H. (ed.): Biogenic structures: their use in interpreting depositional environments, 151–165. *Society of Economic Paleontologists and Mineralogists Special Publication 35*.

Lendzion, K. 1972: Stratigrafia kambru dolnego na obszarze podlasia. *Biuletyn Instytut Geologiczny 233*, 69–157.

Lesquereux, L. 1876: Species of fossil marine plants from the Carboniferous measures. *Annual Report Indiana Geological Survey 7*, 134–135.

Levell, B.K. 1980: Evidence for currents associated with waves in Late Precambrian shelf deposits from Finnmark, North Norway. *Sedimentology 27*, 153–166.

Liñan, E. 1984: Los icnofosiles de la formacion Torrearboles (¿Precambrico?–Cambrico inferior) en los Alfredededores de fuente de Cantos, Badajoz. *Cauadernos do Laboratoria Xeologico de Laxe 8*, 47–74.

Lindström, M. 1977: White sand crosses in turbated siltstone bed, basal Cambrian, Kinnekulle, Sweden. *Geologica et Palaeontologica 11*, 1–8.

Lindström, M. 1984: Do 'white sand crosses' from the Lower Cambrian of Kinnekulle have Glendonite affinity? *Geologiska Föreningens i Stockholm Förhandlingar 106*, 32

Lindström, M. & Vortisch, W. 1978: Sphalerite from transgressive and regressive sediments of the Cambro-Silurian Supercycle, Baltic Shield. *Bulletin of the Geological Society of Denmark 27*, 47–53.

Linnaeus, C. 1747: *Carl Linnaei Wästgöta-Resa.* Stockholm (several editions).

Linnarsson, J.G.O. 1866: *Om de Siluriska Bildningarne i Mellersta Westergötland. 1.* 24 pp. Nisbeth, Stockholm.

Linnarsson, J.G.O. 1868: Bidrag till Westergötlands Geologi. *Öfversikt af Kongliga Vetenskaps-Akademiens Förhandlingar 1868:1*, 53–62.

Linnarsson, J.G.O. 1869a: Om några försteningar från Vestergötlands sandstenslager. *Öfversikt av Kongliga Vetenskaps-Akademiens Handlingar 1869:3*, 337–357. [Translated as 'On some fossils found in the Eophyton Sandstone at Lugnås in Sweden'. Norstedt, Stockholm, 1869, 16 pp. Also as 'On some fossils found at Lugnås in Sweden', *Geological Magazine 6*, 393–406.]

Linnarsson, J.G.O. 1869b: Om Vestergötlands Cambriska och Siluriska aflagringar. *Kongliga Svenska Vetenskaps-Akademiens Handlingar 8:2.* 58 pp.

Linnarsson, J.G.O. 1871: Geognostiska och paleontologiska iakttagelser öfver Eophytonsandstenen i Vestergötland. *Kongliga Svenska Vetenskaps-Akademiens Handlingar 9:7*, 19 pp.

Linnarsson, J.G.O. 1874: Jemförelse mellan den kambrisk-siluriska lagerföljden i Sverige, Böhmen och Ryska Östersjöprovinserna. *Förhandlingar vid det 11te Skandinaviska Naturforskarmötet,* p. 269. København.

Lisson, C.I. 1904: Los *Tigillites* del salto del Fraile y algunes Sonneratoa del Morro Solar. *Boletin del Cuerpo de Ingenieros de Minas del Peru 17.* 64pp.

Ludwig, R. 1869: Fossile Pflanzenresten aus den paläolithischen Formationen der Umgebung von Dillenburg, Biedenkopf und Friedberg und aus dem Saalfeldischen. *Palaeontographica 17*, 105–128.

Lugn, A.L. 1941: The origin of *Daemonelix. Journal of Geology 49*, 673–696.

Lundgren, S.A.B. 1891: Studier öfver fossilförande lösa block. *Geologiska Föreningens i Stockholm Förhandlingar 13*, 111–121.

McCann, T. & Pickerill, R.K. 1988: Flysch trace fossils from the Cretaceous Kodiak Formation of Alaska. *Journal of Paleontology 62*, 330–348.

McCarthy, B. 1979: Trace fossils from a Permian shoreface–foreshore environment, eastern Australia. *Journal of Paleontology 53*, 345–366.

McMenamin, M.A.S. & Schulte-McMenamin, D.L. 1990: *The Emergence of Animals: the Cambrian Breakthrough.* Columbia University Press, New York. 217 pp.

MacNaughton, R.B. & Pickerill, R.K. 1995: Invertebrate ichnology of the nonmarine Lepreau Formation (Triassic) southern New Brunswick, eastern Canada. *Journal of Paleontology 69*, 160–171

Magwood, J.P.A. 1992: Ichnotaxonomy: a burrow by any other name...? *In* Maples, C.G. & West, R.R. *Trace Fossils,* 15–33. *Short Courses in Paleontology 5.*

Magwood, J.P.A. & Pemberton, S.G. 1988: Trace fossils of the Gog Group, a lower Cambrian tidal sand body. *In* Landing, E., Narbonne, G.M. & Myrow, P. (eds.): Trace fossils, small shelly fossils and the Precambrian–Cambrian boundary – proceedings, p. 14. *New York State Museum Bulletin 463.*

Magwood, J.P.A. & Pemberton, S.G. 1990: Stratigraphic significance of *Cruziana:* New data concerning the Cambrian–Ordovician ichnostratigraphic paradigm. *Geology 18*, 729–732.

Manca, V. 1986: Caracteres icnologicos de la Formacion Campanario (Cambrico Superior) en Salta y Jujuy. *Ameghiniana 23*, 75–87.

Maples, C.G. & Archer, A.W. 1987: Redescription of early Pennsylvanian trace-fossil holotypes from the nonmarine Hindostan Whetstone beds of Indiana. *Journal of Paleontology 61*, 890–897.

Martinsson, A. 1965: Aspects of a middle Cambrian thanatotope on Öland. *Geologiska Föreningens i Stockholm Förhandlingar 87*, 81–230.

Martinsson, A. 1970: Toponomy of trace fossils. *In* Crimes, T.P. & Harper, J.C. (eds.): *Trace Fossils,* 323–330. *Geological Journal Special Issue 3.*

Martinsson, A. 1974: The Cambrian of Norden. *In* Holland, C.H. (ed.): *Cambrian of the British Isles, Norden, and Spitsbergen,* 186–282. Wiley, New York, N.Y.

Matthew, G.F. 1891: Illustrations of the fauna of the St. John Group, No. V. *Royal Society of Canada Transactions, Section 4:8*, 123–166.

Matthew, G.F. 1901: *Monocraterion* and *Oldhamia. The Irish Naturalist 10*, 135–136.

Mayer, G. 1954: Neue Beobachtungen an Lebensspuren aus dem unteren Hauptmuschelkalk (Trochitenkalk) von Wiesloch. *Neues Jahrbuch für Geologie und Paläontologie, Abhandlungen 99*, 223–229.

Mayoral, E. 1986: *Gyrolithes vidali* nov. icnoesp. (Plioceno marino) en el sector suroccidental de la Cuneca del Guadalquivir (area de palos de la frontera, Huelva, España). *Estudios geologicos 42*, 211–233.

Miller, M.F. & Johnson, K.G. 1981: *Spirophyton* in alluvial–tidal facies of the Catskill deltaic complex: possible biological control of ichnofossil distribution. *Journal of Paleontology 55*, 1016–1027.

Miller, M.F. & Rehmer, J. 1982: Using biogenic structures to interpret sharp lithologic boundaries: an example from the Lower Devonian of New York. *Journal of Sedimentary Petrology 52*, 887–895.

Miller, S.A. 1879: Description of twelve new fossil species and remarks upon others. *Journal of the Cincinnati Society of Natural History 2*, 104–118.

Miller, S.A. 1889: *North American Geology and Palaeontology for the Use of Amateurs, Students, and Scientists.* 718 pp. Press of Western Methodist Book Concern, Cincinnati.

Miller, S.A. & Dyer, C.B. 1878: Contributions to palaeontology. *Journal of the Cincinnati Society of Natural History 1*, 24–39.

Miller, S.A. & Dyer, C.B. 1878: *Contributions to Palaeontology 2.* 11 pp. Barclay, Cincinnatti, Ohio.

Moberg, J.C. 1911: Historical–stratigraphical review of the Silurian of Sweden. *Sveriges Geologiska Undersökning C 229.* 210 pp.

Moczydłowska, M. 1989: Upper Proterozoic and Lower Cambrian acritarchs from Poland – micropalaeontology, biostratigraphy and thermal study. *Lund Publications in Geology 75.* 30 pp.

Moczydłowska, M. 1991: Acritarch biostratigraphy of the Lower Cambrian and the Precambrian–Cambrian boundary in southeastern Poland. *Fossils and Strata 29*, 127 pp.

Moczydłowska, M. & Vidal, G. 1986: Lower Cambrian acritarch zonation in southern Scandinavia and southeastern Poland. *Geologiska Föreningens i Stockholm Förhandlingar 108*, 201–223.

Moczydłowska, M. & Vidal, G. 1988: How old is the Tommotian? *Geology 16*, 166–168.

[Möller, M. 1987: Die Geologie der Altpaläozoischen Sedimentite des Westlichen Kinnekulle. Unpublished Thesis. Fachbereich Geowissenschaften. Philipps-Universität Marburg/Lahn. 155 pp.]

Moussa, M.T. 1970: Nematode fossil trails from the Green River Formation (Eocene) in the Uinta Basin. *Journal of Paleontology 44*, 304–307.

Myers, A.C. 1977: Sediment processing in a marine subtidal sandy bottom community: II. Biological consequences. *Journal of Marine Research 35*, 633–647.

Myrow, P.M. 1992: Pot and gutter casts from the Chapel Island Formation, southeast Newfoundland. *Journal of Sedimentary Petrology 62*, 992–1007.

Narbonne, G.M. 1984: Trace fossils in Upper Silurian tidal flat to basin slope carbonates of Arctic Canada. *Journal of Paleontology 58*, 398–415.

Narbonne, G.M. & Aitken, J.D. 1990: Ediacaran fossils from the Sekwi Brook area, Mackenzie Mountains, Northwestern Canada. *Palaeontology 33*, 945–980.

Narbonne G.M. & Myrow, P.M. 1988: Trace fossil biostratigraphy in the Precambrian–Cambrian boundary interval. *In* Landing, E., Narbonne, G.M. & Myrow, P. (eds.): Trace fossils, small shelly fossils and the Precambrian–Cambrian boundary, Proceedings, 72–76. *New York State Museum Bulletin 463.*

Narbonne, G.M., Myrow, P.M., Landing, E. & Anderson, M.M. 1987: A candidate stratotype for the Precambrian–Cambrian boundary, Fortune Head, Burin Peninsula, southeastern Newfoundland. *Canadian Journal of Earth Sciences 24*, 1277–1293.

Nathorst, A.G. 1874: Om några förmodade växtfossil. *Öfversikt af Kongliga Vetenskaps-Akademiens Förhandlingar 1873:9*, 25–52.

Nathorst, A.G. 1881a: Om aftryck af Medusor i Sveriges kambriska lager. *Kongliga Svenska Vetenskaps-Akademiens Handlingar 19:1.* 34 pp.

Nathorst, A.G. 1881b: Om spår av några evertebrerade djur m.m. och deras paleontologiska betydelse. *Kongliga Svenska Vetenskaps-Akademiens Handlingar 18:7.* 59 pp.

Nathorst, A.G. 1885: Om kambriska pyramidalstenar. *Öfversigt af Kongliga Vetenskaps-Akademiens Förhandlingar 1885:10*, 5–17.

Nathorst, A.G. 1886a: Nouvelles observations sur des traces d'Animaux et autres phenomenes d'origine purement mecanique decrits commes 'Algues Fossiles'.*Kongliga Svenska Vetenskaps-Akademiens Handlingar 21:14.* 58 pp.

Nathorst, A.G. 1886b: Om de sandslipade stenarnes förekomst i de kambriska lagren vid Lugnås. *Öfversigt av Kongliga Svenska Vetenskaps-Akademiens Förhandlingar 1886:6*, 185–192.

Nathorst, A.G. 1910: Ein besonders instruktives Exemplar unter den Medusenabdrucken aus dem kambrischen Sandstein bei Lugnås. *Sveriges Geologiska Undersökning C 228*, 1–9.

Nowlan, G.S., Narbonne, G.M. & Fritz, W.W. 1985: Small shelly fossils and trace fossils near the Precambrian–Cambrian boundary in the Yukon Territory, Canada. *Lethaia 18*, 233–256.

Öpik, A.A. 1925: Beitrag zur Stratigraphie und Fauna des estnischen Unter-Kambriums (Eophyton-Sandstein). *Publications of the Geological Institution of the University of Tartu 3.* 20 pp.

Öpik, A.A. 1933: Über *Scolithus* aus Estland. *Acta et Commentationes Universitatis Tartuensis (Dorpatensis), Ser A, 24:3.* 12 pp.

Öpik, A.A. 1956: Cambrian (Lower Cambrian) of Estonia. *In* Rodgers, J. (ed.): *XX Congreso Geologico Internacional. El Sistema Cambrico su Paleografía y el problema de su base, Part I*, 97–126.

Orłowski, S. 1974: Lower Cambrian biostratigraphy in the Holy Cross Mts., based on the trilobite family Olenellidae. *Acta Geologica Polonica 24*, 1–16.

Orłowski, S. 1989: Trace fossils in the Lower Cambrian sequence in the Świętokrzyskie Mountains, central Poland. *Acta Palaeontologica Polonica 34*, 211–231.

Orłowski, S. 1992: Trilobite trace fossils and their stratigraphical significance in the Cambrian sequence of the Holy Cross Mountains, Poland. *Geological Journal 27*, 15–35.

Orłowski, S. & Radwański, A. 1986: Middle Devonian sea-anemone burrows, *Alpertia santacrucensis* ichnogen. et ichnosp. n., from the Holy Cross Mountains. *Acta Geologica Polonica 36*, 233–249.

Orłowski, S., Radwański, A. & Roniewicz, P. 1970: The trilobite ichnocoenoses in the Cambrian sequence of the Holy Cross Mountains. *In* Crimes, T.P. & Harper, J.C. (eds.): *Trace Fossils. Geological Journal Special Issue 3*, 345–360.

Orłowski, S., Radwański, A. & Roniewicz, P. 1971: Ichnospecific variability of the Upper Cambrian *Rusophycus* from the Holy Cross Mts. *Acta Geologica Polonica 21*, 341–348.

Osgood, R.G. 1970: Trace fossils of the Cincinnati area. *Palaeontographica Americana 6:41*, 281–444.

Osgood, R.G. & Drennen, W.T. 1975: Trilobite trace fossils from the Clinton Group (Silurian) of east-central New York State. *Bulletins of American Paleontology 67:287*, 300–348.

Pacześna, J. 1985: Skamieniałości śladowne górnego wendu i dolnego kambru południowej Lubelszczyzny. *Kwartalnik Geologiczny 29*, 255–270.

Pacześna, J. 1986: Upper Vendian and Lower Cambrian ichnocoenoses of Lublin Region. *Biuletyn Instytutu Geologicznego 355*, 31–47.

Pacześna, J. 1989: Polski i globalny zapis biodarzenia na granicy Prekambr–Kambr. *Przeglad Geologiczny 29*, 542–546.

Palij, V.M. 1976: Ostatki besskeletnoj fauny i sledy zhiznedeyatel'nosti iz otlozhenij verkhnego dokembriya i nizhnego kembriya Podolii. [Remains of soft-bodied fauna and trace fossils from the upper Precambrian and lower Cambrian of Podolia.] *In: Paleontologiya i Stratigrafiya Verkhnego Dokembriya i Nizhnego Paleozoya Jugo-Zapadna Vostochno-Evropejskoj Platformy*, 63–77. Naukova Dumka, Kiev.

Palij, V.M., Posti, E. & Fedonkin, M.A. 1983: Soft-bodied Metazoa and animal trace fossils in the Vendian and early Cambrian. *In* Urbanek, A. & Rozanov, A. Yu. (eds.): *Upper Precambrian and Cambrian Palaeontology of the East-European Platform*, 56–94. Wydawnictea Geologiczne, Warszawa.

Pek, I. & Gába, Z. 1983: *Syringomorpha nilssoni* (Torell) z ledovcových uloženin severní Moravy (ČSSR). *Acta Universitas Palackianae Olomucensis Facultas Rerum Naturalium 77, Geographica–Geologica 22*, 21–28.

Pemberton, S.G. & Frey, W.F. 1982: Trace fossil nomenclature and the *Planolites–Palaeophycus* dilemma. *Journal of Paleontology 56*, 843–881.

Pemberton, S.G. & Magwood, J.P.A. 1990: A unique occurrence of *Bergaueria* in the Lower Cambrian Gog Group near Lake Loise, Alberta. *Journal of Paleontology 64*, 436–440.

Pemberton, S.G., Frey, R.W. & Bromley, R.G. 1988: The ichnotaxonomy of *Conostichus* and other plug-shaped ichnofossils. *Canadian Journal of Earth Sciences 25*, 866–892.

Péneau, J: 1946: Étude sur l'Ordovicien inférieur (Arénigien = Grès armoricain) et sa Faune (spécialment en Anjou). *Bulletin de la Société d'Études Scientifiques d'Angers, Nouvelles Série LXXIVe à LXXVIes Années 1944–1946*, 37–106.

Pickerill, R.K. 1977: Trace fossils from the Upper Ordovician (Caradoc) of the Berwyn Hills, Central Wales. *Geological Journal 12*, 1–16.

Pickerill, R.K. 1980: Phanerozoic flysch trace fossil diversity – observations based on an Ordovician flysch ichnofauna from the Aroostook–Matapedia Carbonate Belt of northern New Brunswick. *Canadian Journal of Earth Sciences 17*, 1259–1270.

Pickerill, R.K. 1989: *Bergaueria perata* Prantl, 1945 from the Silurian of Cape George, Nova Scotia. *Atlantic Geology 25*, 191–197.

Pickerill, R.K. 1994: Nomenclature and taxonomy of invertebrate trace fossils. *In* Donovan, S.K. (ed.): *The Palaeobiology of Trace Fossils*, 3–42. Wiley, Chichester.

Pickerill, R.K. & Narbonne, G.M. 1995: Composite and compound ichnotaxa: a case example from the Ordovician of Quèbec, eastern Canada. *Ichnos 4*, 53–69.

Pickerill, R.K. & Peel, J.S. 1990: Trace fossils from the Lower Cambrian Bastion Formation of North-East Greenland. *In* Peel, J.S. (ed.): Lower Cambrian trace fossils from Greenland. *Grønlands Geologiske Undersøgelse, Rapport 147*, 5–43.

Pickerill, R.K., Fyffe, L.R. & Forbes, W.H. 1987: Late Ordovician–Early Silurian trace fossils from the Matapedia Group, Tobique River, western New Brunswick, Canada. *Maritime Sediments and Atlanic Geology 23*, 77–88.

Pickerill, R.K., Fyffe, L.R. & Forbes, W.H. 1988: Late Ordovician–Early Silurian trace fossils from the Matapedia Group, Tobique River, western New Brunswick, Canada. II. Additional discoveries with descriptions and comments. *Maritime Sediments and Atlanic Geology 24*, 139–148.

Piénkowski, G. & Westwalewicz-Mogilska, E. 1986: Trace fossils from the Podhale Flysch Basin, Poland – an example of ecologically based lithocorrelation. *Lethaia 19*, 53–65.

Piper, J.D. 1985: Continental movements and breakup in Late Precambrian–Cambrian times: prelude to Caledonian orogenesis. *In* Gee,

D.G. & Sturt, B.A. (eds.): *The Caledonide Orogen – Scandinavia and Related Areas*, 19–34. Wiley, New York, N.Y.

Plička, M. 1968: *Zoophycos*, and a proposed classification of sabellid worms. *Journal of Paleontology* 42, 836–849.

Plummer, P.S. & Gostin, V.A. 1981: Shrinkage cracks: desiccation or synaeresis? *Journal of Sedimentary Petrology* 51, 1147–1156.

Pollard, J.E. 1981: A comparison between the Triassic trace fossils of Cheshire and south Germany. *Palaeontology* 24, 555–588.

Pollard, J.E. 1985: *Isopodichnus*, related arthropod trace fossils and notostracans from Triassic fluvial sediments. *Transactions of the Royal Society of Edinburgh, Earth Sciences* 76, 273–285.

Poulsen, C. 1967: Fossils from the Lower Cambrian of Bornholm. *Danske Videnskabernes Selskab, Matematisk-Fysiske Meddelelser* 36:2, 1–48.

Powell, E.N. 1977: The relationship of the trace fossil *Gyrolithes* (*Xenohelix*) to the family Capitellidae (Polychaeta). *Journal of Paleontology* 51, 552–556.

Prantl, F. 1946: Dvě záhadné zkameněliny (stopy) z vrstev chrustenických – d83. *Rozpravy České Akademie Věd a Umění, Třída II*, 55:3, 1–18.

Raaf, J.F.M. de, Boersma, J.R. & van Gelder, A. 1977. Wave-generated structures and sequences from a shallow marine succession, Lower Carboniferous, County Cork, Ireland. *Sedimentology* 24, 451-483.

Radwański, A. & Roniewicz, P. 1960: Ripple marks and other sedimentary structures of the Upper Cambrian at Wielka Wisniówka. *Acta Geologica Polonica* 10, 371–399.

Reif, W.-E. 1982: Muschelkalk/Keuper Bone-Beds (Middle Triassic, SW-Germany) – Storm condensation in a regressive cycle. *In* Einsele, G. & Seilacher, A. (eds.): *Cyclic and Event Stratification*, 299–325. Springer, Berlin.

Reineck, H.-E. 1958: Wühlbau-Gefüge in Abhängigkeit von Sediment-Umlagerungen. *Senckenbergiana Lethaea* 39, 1–23.

Reineck, H.E. 1969: Die Entstehung von Runzelmarken. *Natur und Museum* 99, 386–388

Reineck, H.E., Gutmann, W.F. & Hertweck, G. 1967: Das Schlickgebiet südlich Helgoland als Beispiel rezenter Schelfablagerung. *Senckenbergiana Lethaea* 48, 219–275.

Reineck, H.E. & Singh, I.B. 1980: *Depositional Sedimentary Environments with Reference to Terrigenous Clastics*. 2 ed. 549 pp. Springer, Berlin.

Ricci Lucchi, F. 1970: *Sedimentografia. Atlante Fotografico Delle Strutture Primarie dei Sedimenti*. 288 pp. Zanichelli, Bologna.

Richardson, J.D. 1975: Trace fossils from the Wolfe City Formation (Upper Cretaceous) Northern Texas. *The Texas Journal of Science* 26, 339–352.

Richter, D. 1971: Ballen und Kissen (ball-and-pillow structure), eine weitverbreitete, bisher wenig bekannte Sedimentstruktur. *Forschungsberichte des landes Nordrhein–Westfalen* 2184. 47 pp.

Richter, R. 1927: *Syringomorpha nilssoni* (Torell) in nord-deutschen Geschieben das schwedischen Cambriums. *Senckenbergiana* 9, 260–268.

Richter, R. 1937: Marken und Spuren aus allen Zeiten. I–II. I. Wühl-Gefüge durch kot-gefüllte Tunnel (*Planolites montanus* n.sp.) aus dem Ober-Karbon der Ruhr. *Senckenbergiana* 19, 151–159.

Rijken, M. 1979: Food and food uptake in *Arenicola marina. Netherlands Journal of Sea Research* 13, 406–421.

Rindsberg, A.K. & Gastaldo, R.A. 1990: New insights on ichnogenus *Rosselia* (Cretaceous and Holocene, Alabama). *Journal of the Alabama Academy of Science* 61, 154.

Romano, M. & Melendez, B. 1985: An arthropod (merostome) ichnocoenosis from the Carboniferous of northwest Spain. *In* Dutro, J.T. & Pfefferkorn, H.W. (eds.): *Neuvième Congrès International de Stratigraphie et de Géologie du Carbonifère. Washington and Champaign–Urbana. Compte rendu* 5, 317–325. Southern Illinois University Press, Carbondale and Edwardsville, Ill.

Romano, M. & Whyte, M.A. 1987: A limulid trace fossil from the Scarborough Formation (Jurassic) of Yorkshire; its occurrence, taxonomy and interpretation. *Proceedings of the Yorkshire Geological Society* 46, 85–95.

Romano, M. & Whyte, M.A. 1990: *Selenichnites*, a new name for the ichnogenus *Selenichnus* Romano & Whyte, 1987. *Proceedings of the Yorkshire Geological Society* 48, 221.

Roniewicz, P. & Pienkowski, G. 1977: Trace fossils of the Podhale Flysch Basin. *In* Crimes, T.P. & Harper, J.C. (eds.): *Trace Fossils 2. Geological Journal Special Issue* 2, 273–288.

Rouault, M. 1851: Note préliminaire sur une nouvelle formation découverte dans la terrain silurien inférieur de la Bretagne. *Bulletin Société Géologique de France, Sér.* 2, 7, 724–744.

Runnegar, B. 1992: Evolution of the earliest animals. *In* Schopf, J.W. (ed.): *Major Events in the History of Life*, 65–93. Jones and Bartlett, Boston, Mass.

Salter, J.W. 1853: On the lowest fossiliferous beds of north Wales. *British Association for the Advancement of Science, Report for 1852*, 56–68.

Saporta, L.C.J.G. de 1884: *Les organismes problematiques des anciennes mers*. 102 pp. Masson, Paris.

Shaffer, F.X. 1928: *Hormosiroidea florentina* n.g., n.sp., ein Fucus aus der Kreide der Umgebung von Florenz. *Paläontologisches Zeitschrift* 10, 212–215.

Schindewolf, O.H. 1921: Studien aus dem Marburger Buntsandstein 1, 2. *Senckenbergiana* 3, 33–49.

Schindewolf, O.H. 1928: Studien aus dem Marburger Buntsandstein. IV. *Isopodichnus problematicus* (SCHDWF.) im Unteren und Mittleren Buntsandstein. *Senckenbergiana* 10, 27–37.

Schmalfuss, H. 1981: Structure, patterns and function of cuticular terrases in trilobites. *Lethaia* 14, 331–341.

Schmidt, F. 1888: Über eine neuentdeckte untercambrische Fauna in Estland. *Mémoires de l'Academie Impériale des Sciences de St-Pètersbourg* 7, 36:2, 1–27.

Scholz, H. 1977: Bemerkungen zum Mickwitziasandstein. *Mitteilungen des Naturwissenschaftlichen Arbeitskreises Kempten* 21, 61–75.

Seilacher, A. 1953: Studien zur Palichnologie II. Die fossilen Ruhespuren (Cubichnia). *Neues Jahrbuch für Geologie und Paläontologie, Abhandlungen* 98, 87–124.

Seilacher, A. 1955a: Spuren und Lebensweise der Trilobiten. *In* Schindewolf, O.H. & Seilacher, A. Beiträge zur Kenntniss des Kambriums in der Salt Range (Pakistan). *Akademie der Wissenschaften und der Literatur. Abhandlungen der Mathematisch-Naturwissenschaftlichen Klasse* 1955, 342–372.

Seilacher, A. 1955b: Spuren und Fazies im Unterkambrium. *In* Schindewolf, O.H. & Seilacher, A. Beiträge zur Kenntniss des Kambriums in der Salt Range (Pakistan), *Akademie der Wissenschaften und der Literatur, Mainz. Abhandlungen der Mathematisch-Naturwissenschaftlichen Klasse* 1955, 373–399.

Seilacher, A. 1956: Der Beginn des Kambriums als biologische Wende. *Neues Jahrbuch für Geologie und Paläontologie, Abhandlungen* 103, 155–180.

Seilacher, A. 1959a: Zur ökologischen Charakteristik von Flysch und Molasse. *Eclogae Geologicae Helvetiae* 51, 1062–1078.

Seilacher, A. 1959b: Vom Leben der Trilobiten. *Die Naturwissenschaften* 46, 389–393.

Seilacher, A. 1960: Lebenspuren als Leitfossilien. *Geologische Rundschau* 49, 41–50.

Seilacher, A. 1962: Form und Funktion des Trilobiten-Daktylus. *Paläontologische Zeitschrift. Festband Hermann Schmidt*, 218–227.

Seilacher, A. 1963: Lebenspuren und Salinitätsfazies. *Fortschritte in der Geologie von Rheinland und Westfalens* 10, 81–94.

Seilacher, A. 1964: Biogenic sedimentary structures. *In* Imbrie, J. & Newell, N. (eds.): *Approaches to Paleoecology*, 296–316. Wiley, New York, N.Y.

Seilacher, A. 1967: Bathymetry of trace fossils. *Marine Geology* 5, 413–428.

Seilacher, A. 1970: *Cruziana* stratigraphy of nonfossiliferous Paleozoic sandstones. *In* Crimes, T.P. & Harper, J.C. (eds.): *Trace Fossils*, 447–476. *Geological Journal Special Issue* 3.

Seilacher, A. 1977: Pattern analysis of *Paleodictyon* and related trace fossils. *In* Crimes, T.P. & Harper, J.C. (eds.): *Trace fossils 2*, 289–334. *Geological Journal Special Issue* 9.

Seilacher, A. 1982: Distinctive features of sandy tempestites. *In* Einsele, G. & Seilacher, A. (eds.): *Cyclic and Event Stratification*, 333–349. Springer, Berlin.

Seilacher, A. 1983a: Paleozoic sandstones in southern Jordan: Trace fossils, depositional environments and biogeography. *In* Abed, A.M. & Khaled, H.M. (eds.): *Geology of Jordan. Proceedings of the First Jordanian Geological Conference*, 209–222. Jordanian Geologists Association, Amman.

Seilacher, A. 1983b: Upper Paleozoic trace fossils from the Gilf Kebir–Abu Ras area in southwestern Egypt. *Journal of African Earth Sciences 1*, 21–34.

Seilacher, A. 1984: Late Precambrian and Early Cambrian Metazoa: Preservational or Real Extinctions? *In* Holland, H.D. & Trendall, A.F. (eds.): *Patterns of Change in Earth Evolution*, 159–168. Springer-Verlag, Berlin.

Seilacher, A. 1985: Trilobite palaeobiology and substrate relationship. *Transactions of the Royal Society of Edinburgh 76*, 231–237.

Seilacher, A. 1986: Evolution of behavior as expressed in marine trace fossils. *In* Nitecki, M.H. & Kitchell, J.A. (eds.): *Evolution of Animal Behavior. Paleontological and Field Approaches*, 63–87. Oxford University Press, New York.

Seilacher, A. 1990: Paleozoic trace fossils. *In* Said, R. (ed.): *The Geology of Egypt*, 649–670. Balkema. Rotterdam.

Seilacher, A. 1992a: Vendobionta and Psammocorallia: lost constructions of Precambrian evolution. *Journal of the Geological Society, London 149*, 607–613.

Seilacher, A. 1992b: An updated *Cruziana* stratigraphy of Gondwanan Paleozoic sandstones. *In* Salem, M.J. & Busrewil, M.T. (eds.): *The Geology of Libya*, 1565–1581. Elsevier, Amsterdam.

Seilacher, A. 1994a: Early multicellular life: Late Proterozoic fossils and the Cambrian explosion. *In* Bengtson, S. (ed.): *Early Life on Earth. Nobel Symposium No. 84*, 389–400. Columbia University Press, New York, N.Y.

Seilacher, A. 1994b: How valid is *Cruziana* stratigraphy? *Geologisches Rundschau 83*, 752–758.

Seilacher, A. & Crimes, T.P. 1969: 'European' species of trilobite burrows in eastern Newfoundland. *In* Kay, M. (ed.): *North Atlantic – Geology and Continental Drift*, 145–148. American Association of Petroleum Geologists Memoir 12.

Seilacher, A. & Hemleben, C. 1966: Beiträge zur Sedimentation und Fossilführung des Hunsrückschiefers 14. Spurenfauna und Bildungstiefe der Hunsrückschiefer (Unterdevon). *Notizblatt des Hessischen Landesamtes für Bodenforschung zu Wiesbaden 94*, 40–53.

Seilacher, A. & Meischner, D. 1965: Fazies-Analyse im Paläozoikum des Oslo-Gebietes. *Geologische Rundschau 54*, 596–619.

Sepkoski, J.J. 1987: Environmental trends in extinction during the Paleozoic. *Science 235*, 64–66.

Signor, P.W. 1994: Proterozoic–Cambrian boundary trace fossils: Biostratigraphic significance of *Harlaniella* in the Lower Cambrian Wood Canyon Formation, Death Valley, California. *New York State Museum Bulletin 481*, 317–322.

Simpson, S. 1970: Notes on *Zoophycos* and *Spirophyton*. *In* Crimes, T.P. & Harper, J.C. (eds.): *Trace Fossils*, 505–514. *Geological Journal Special Issue 3.*

Singh, I.B. & Rai, V. 1983: Fauna and biogenic structures in Krol–Tal succession (Vendian–early Cambrian), lesser Himalaya: their biostratigraphic and palaeoecological significance. *Journal of the Palaeontological Society of India 28*, 67–90.

Singh, I.B. & Wunderlich, F. 1978: On the terms wrinkle marks (Runzelmarken), millimetre ripples, and mini ripples. *Senckenbergiana Maritima 10*, 75–83.

Smith, R.M.H. 1987: Helical burrow casts of therapsid origin from the Beaufort Group (Permian) of South Africa. *Palaeogeography, Palaeoclimatology, Palaeoecology 60*, 155–170.

Stanley, D.C.A. & Pickerill, R.K. 1993: *Fustiglyphus annulatus* from the Ordovician of Ontario, Canada, with a systematic review of the ichnogenera *Fustiglyphs* Vialov 1971 and *Rhabdoglyphus* Vassoievich 1951. *Ichnos 3*, 57–67.

Steimle, F.W. & Sindemann, C.J. 1978. Review of oxygen depletion and associated mass mortalities in the Middle Atlantic Bight in 1976. *Marine Fisheries Review 40*, 17–26.

Størmer, L. 1956: A Lower Cambrian merostome from Sweden. *Arkiv för Zoologi 9*, 507–517.

Taylor, B.J. 1967: Trace fossils from the Fossil Bluff Series of Alexander Island. *British Antarctic Survey, Bulletin 13*, 1–30.

Tedesco, L.E. & Wanless, H.R. 1991: Generation of sedimentary fabrics and facies by repetitive excavation and storm infilling of burrow networks, Holocene of South Florida and Caicos Platform, B.W.I. *Palaios 6*, 326–343.

Thayer, C.W. 1983: Sediment-mediated biological disturbance and the evolution of marine benthos. *In* Tevesz, M.J.S. & McCall, P.L. (eds.): *Biotic Interactions in Recent and Fossil Benthic Communities*, 479–625. Plenum, New York, N.Y.

Thorslund, P. 1958: Djupborrningen på Gotska Sandön. *Geologiska Föreningens i Stockholm Förhandlingar 80*, 190–197.

Thorslund, P. 1960: The Cambro-Silurian of Sweden. *In* Magnusson, N.H., Thorslund, P., Brotzen, F., Asklund, B. & Kulling, O.: Description to accompany the map of the pre-Quaternary rocks of Sweden, 96–110. *Sveriges Geologiska Undersökning Ba 16.*

Thorslund, P. & Axberg, S. 1979: Geology of the southern Bothnian Sea. Part I. *Bulletin of the Geological Institutions of the University of Uppsala, N.S. 8*, 35–62.

Thorslund, P. & Westergård, A.H. 1938: Deep boring through the Cambro-Silurian at File Haidar, Gotland. *Sveriges Geologiska Undersökning C 415*, 1–56.

Toots, H. 1963: Helical burrows as fossil movement patterns. *Contributions to Geology 2*, 129–134.

Torell, O. 1868: Bidrag till Sparagmitens geognosi och paleontologi. *Lunds Universitets Årsskrift 4, Afd 2:3*, 1–40.

Torell, O. 1870: Petrificata Suecana Formationis Cambricae. *Lunds Universitets Årsskrift 6, Afd 2:8*, 1–14.

Tromelin, G. de & Lebesconte, P. 1876: Essai d'un catalogue raisonné des fossiles siluriens des départements de Main-et-Loire, de la Loire-Inférieure et du Morbihan. *Association Francaise pour l'Avancement des Sciences, Compte Rendus 4*, 601–661.

Uchman, A. 1995: Taxonomy and palaeoecology of flysch trace fossils: The Marnoso-arenacea Formation and associated facies (Miocene, Northern Apennines, Italy). *Beringeria 15*, 115 pp.

Uchman, A. & Krenmayer, H.G. 1995: Trace fossils from the Lower Miocene (Ottnangian) molasse deposits of Upper Austria. *Paläontologische Zeithschrift 69*, 503–524.

Vanuxem, L. 1842: *Natural History of New-York. Geology of New-York Part III. Comprising the Survey of the Third Geological District.* 306 pp. White & Visscher, Albany.

Vassoevich, N.B. 1932: O nekotorykh priznakakh pozvolyayushchikh otlochetoprokinutoe polozhenie flishevykh obrazovaniy ot normalynogo. *Trudy Geologicheskogo Instituta AN SSSR 2*, 47–63.

Vialov, O.S. 1969: Vintoobraznyj khod chlenistonogogo iz Kryma. [Spiral arthropod traces from Crimea.] *Paleontologicheskij Sbornik, Lvov 6:1*, 105–109.

Vialov, O.S. 1971. Redkie problematiki iz mesozoya Pamira i Kavkaza, *Paleontologicheskij Sbornik, Lvov, 7:2*, 85–93.

Vidal, G. 1981a: Lower Cambrian acritarch stratigraphy in Scandinavia. *Geologiska Föreningens i Stockholm Förhandlingar 103*, 183–192.

Vidal, G. 1981b: Micropaleontology and biostratigraphy of the Lower Cambrian sequence in Scandinavia. *In* Taylor, M.E. (ed.): *Short papers for the Second International Symposium on the Cambrian System 1981*, 232–235. *U.S. Geological Survey Open-FileReport 81-743.*

Vogt, T. 1924: Forholdet mellem sparagmitsystemet og det marine underkambrium vid Mjøsen. *Norsk Geologisk Tidskrift 7*, 281–384.

Volk, M.1968: *Trichophycus thuringicum*, eine Lebensspuren aus den Phycoden-Schichten (Ordovizium) Thüringens. *Senckenbergiana Lethaea 49*, 581–585.

Waern, B. 1952: Palaeontology and stratigraphy of the Cambrian and lowermost Ordovician of the Bödahamn core. *Bulletin of the Geological Institutions of the University of Upsala 34*, 223–250.

Walcott, C.D. 1890: The fauna of the Lower Cambrian or Olenellus Zone. *United States Geological Survey Annual Report 10:1*, 509–774.

Walcott, C.D. 1896: Fossil jelly fishes from the Middle Cambrian terrane. *Proceedings of the United States National Museum 42*, 611–614.

Walcott, C.D. 1898: Fossil Medusae. *United States Geological Survey, Monographs 30.* 201 pp.

Walcott, C.D. 1914: Cambrian geology and paleontology 3, No. 2. Pre-Cambrian Algonkian algal flora. *Smithsonian Miscellaneous Collections 64:2*, 77–156.

Walcott, C.D. 1918: Cambrian geology and paleontology 4, No. 4. Appendages of trilobites. *Smithsonian Miscellaneous Collections 67:4*, 115–216.

Wallin, J.A. 1868: *Bidrag till kännedommen om Vestgötabergens byggnad, 1.* 27 pp. Lundbergs, Lund.

Walter, H. 1984: Zur Ichnologie der Arthropoda. *Freiberger Forschungshefte C391*, 58–94.

Walter, M.R., Elphistone, R. & Heys, G.R. 1989: Proterozoic and Early Cambrian trace fossils from the Amadeus and Georgina Basins, central Australia. *Alcheringa 13*, 209–256.

Wanless, H.R. & Tedesco, L.R. & Tyrrell, K.M. 1988: Production of subtidal tubular and surficial tabular tempestites by Hurricane Kate, Caicos Platform, British West Indies. *Journal of Sedimentary Petrology 58*, 739–750.

Webby, B.D. 1970: Late Precambrian trace fossils from New South Wales. *Lethaia 3*, 79–109

Webby, B.D. 1983: Lower Ordovician arthropod trace fossils from western New South Wales. *Proceedings Linnean Society of New South Wales 107*, 59–74.

Weiss, W. 1941: Die Enstehung der 'Zöpfe' im schwarzen und braunen Jura. *Natur und Volk 71*, 179–184.

Westergård, A.H. 1929: A deep boring through Middle and Lower Cambrian strata at Borgholm, Isle of Öland. *Sveriges Geologiska Undersökning C355.* 19 pp.

Westergård, A.H. 1931a: *Diplocraterion, Monocraterion* and *Scolithus. Sveriges Geologiska Undersökning C 372*, 25 pp.

Westergård, A.H. 1931b: Den kambro-silurska lagerserien. *In* Lundquist, G., Högbom, A. & Westergård, A.H. Beskrivning till kartbladet Lugnås, 29–67. *Sveriges Geologiska Undersökning, Aa 172.*

Westergård, A.H. 1943: Den kambro-siluriska lagerserien. *In* Johansson, S., Sundius, N. & Westergård, A.H.: Beskrivning till kartbladet Lidköping, 22–89. *Sveriges Geologiska Undersökning Aa 182.*

Wetzel, A. & Werner, F. 1981: Morphology and ecological significance of *Zoophycos* in deep-sea sediments off NW Africa. *Palaeogeography, Palaeoclimatology, Palaeoecology 32*, 185–212.

Whittington, H.B. 1975: Trilobites with appendages from the Middle Cambrian, Burgess Shale, British Columbia. *Fossils and Strata 4*, 97–136.

Whittington, H.B. 1980: Exoskeleton, moult stage, appendage morphology and habits of Middle Cambrian trilobite *Olenoides serratus. Palaeontology 23*, 171–204.

Whittington, H.B. & Almond, J.E. 1987: Appendages and habits of the Upper Ordovician trilobite *Triarthrus eatoni. Philosophical Transactions of the Royal Society, London B317*, 1–46.

Willmer, P. 1990: *Invertebrate Relationships. Patterns in Animal Evolution.* 400 pp. Cambridge University Press.

Wilson, A.E. 1948: Miscellaneous classes of fossils, Ottawa Formation, Ottawa – St. Lawrence Valley, Canada. *Department of Mines and Resources, Geological Survey Bulletin 11.* 116 pp.

Wiman, C. 1943: Sandabgüsse von Quallen. *Natur und Volk 73*, 118–121.

Yang Shipu, 1984: Silurian trace fossils from the Yangzi gorges and their significance to depositional environments. *Acta Palaeontologica Sinica 23*, 705–715

Yang Shipu 1990: Stratigraphic range and geographic distribution of *Cruziana* in China and its paleoenvironmental significance. *Earth Science 15*, 263–273.

Yang Shipu, Song Zhimin & Liang Dingyi 1986: Trace fossils from Jurassic and Cretaceous flysches in Shiquanhe–Duoma region, western Tibet. *Proceedings of the Symposium on Mesozoic and Cenozoic geology, China*, 219–228.

Young, F.G. 1972: Early Cambrian and older trace fossils from the Southern Cordillera of Canada. *Canadian Journal of Earth Sciences 9*, 1–17.

Young, G.M. 1967: Possible organic structures in early Proterozoic (Huronian) rocks of Ontario. *Canadian Journal of Earth Sciences 4*, 565–568

Young, G.M. 1969: Inorganic origin of corrugated vermiform structures in the Huronian Gordon Lake Formation near Flack Lake, Ontario. *Canadian Journal of Earth Sciences 6*, 795–799.